Advances in Hydrological Forecasting

Advances in Hydrological Forecasting

Editors

Minxue He
Haksu Lee
Sungwook Wi

MDPI • Basel • Beijing • Wuhan • Barcelona • Belgrade • Manchester • Tokyo • Cluj • Tianjin

Editors
Minxue He
California Department of Water Resources
N/A
Sacramento
United States

Haksu Lee
Len Technologies
N/A
Oak Hill
United States

Sungwook Wi
Civil and Environmental Engineering
University of Massachusetts Amherst
Amherst
United States

Editorial Office
MDPI
St. Alban-Anlage 66
4052 Basel, Switzerland

This is a reprint of articles from the Special Issue published online in the open access journal *Forecasting* (ISSN 2571-9394) (available at: www.mdpi.com/journal/forecasting/special_issues/Hydrological_fc).

For citation purposes, cite each article independently as indicated on the article page online and as indicated below:

LastName, A.A.; LastName, B.B.; LastName, C.C. Article Title. *Journal Name* **Year**, *Volume Number*, Page Range.

ISBN 978-3-0365-1680-6 (Hbk)
ISBN 978-3-0365-1679-0 (PDF)

Contents

forecasting

MDPI

Editorial

Advances in Hydrological Forecasting

Minxue He [1],* and Haksu Lee [2]

[1] California Department of Water Resources, 1416 9th Street, Sacramento, CA 95814, USA
[2] Len Technologies, 12139 Westwood Hills Dr, Oak Hill, VA 20171, USA; haksulee@hanmail.net
* Correspondence: minxuehe@gmail.com

Citation: He, M.; Lee, H. Advances in Hydrological Forecasting. *Forecasting* 2021, 3, 517–519. https://doi.org/10.3390/forecast3030032

Received: 6 July 2021
Accepted: 7 July 2021
Published: 8 July 2021

Publisher's Note: MDPI stays neutral with regard to jurisdictional claims in published maps and institutional affiliations.

Hydrological forecasting is of primary importance to better inform decision-making on flood management, drought mitigation, water system operations, water resources planning, and hydropower generation, among others. Typical hydrological forecasting translates single deterministic or an ensemble of short, intermediate, and long lead-time meteorological forecasts into estimates of hydrological variables of interest (e.g., streamflow, river stage, snowmelt, etc.) via forecast models at the corresponding temporal scales. These models range from process-based hydrological models to purely data-driven models. The model predictive skill and uncertainty are normally verified by comparing archived forecasts to field data or in a hindcasting mode. During forecasting, real-time in situ or remote sensing measurements for forecast hydrological variables can be assimilated into the forecast model to update model states or parameters for improved forecasts. Before being disseminated for operational use, hydrological forecasts are often post-processed to best reflect the perceptions of forecasters on the future state of those forecast variables.

Although there has been immense progress in forecasting systems, services, and observational sensors to date, hydrological forecasting today faces convoluted challenges induced by the increasing trend of extreme events, changing basin climate and hydrology, and demands of a unified and versatile hydrological forecasting system operating at local to continental scales. This Special Issue aimed to explore the latest methodological advances and novel applications in hydrological forecasting that tackle one or more of those challenges. The collection of papers published in this Special Issue covered a range of topics on improving hydrological forecasting via new datasets and innovative approaches. Each of these papers is summarized as follows.

Wu [1] proposed an approach using the bivariate meta-Gaussian distribution that allows the explicit treatment of precipitation intermittency, as well as a wide choice of parametric and non-parametric models for marginal distributions. Wu [1] presented the new proof for an intermittency equation of the model, and studied how to tune the bivariate meta-Gaussian distribution for an optimal fit. The two variables used are predicted single-valued precipitation amounts and the corresponding observed precipitation amounts. Numerical simulations are implemented with data from the Global Ensemble Forecast System and four River Forecast Centers in the United States. The Mallows distance for the entire joint distribution is used to tune the parameter of the meta-Gaussian distribution. The optimization results were comparatively evaluated with results from using the sample correlation coefficient and maximum likelihood estimate as parameter values. The results suggested that tuning the dependence parameter has limited effects toward a better overall model fit, and that the goodness-of-fit of the conditional distribution can be improved in the case of optimizing the parameter for a targeted conditional distribution.

Bhuiyan et al. [2] explored the application of innovative machine-learning (ML)-based error models in improving the quality of satellite-based precipitation products. Specifically, the study applied two ML techniques including the Random Forecast (RF) and Neural Network (NN) to correct errors in the Integrated Multi-satellite Retrievals for Global Precipitation Measurement (IMERG) precipitation product. The proposed error models were exemplified in the Brahmaputra River Basin during a four-year period. The quantile–quantile plot, along with a set of statistical metrics, were utilized to assess the performance of the error models. The findings indicated that both RF-based and NN-based error models lead to reduction in the random and systematic errors for all precipitation percentile ranges. The error corrected IMERG precipitation products showed a slightly higher improvement by the NN-based error model compared to the RF-based model. Overall, the NN-based error model slightly outperformed the RF-based error model. The authors expected that the ML-based error modeling algorithm is potentially applicable to ungauged areas or to the global scale, as well as for other types of remote-sensing precipitation products.

Pokhrel et al. [3] presented a framework to assess future flood rates and risks. The framework consists of a statistical analysis procedure, as well as a hydraulic modeling and risk assessment procedure. The former first bias-corrects streamflow projections based on the latest multi-model ensemble Coupled Model Inter-comparison Project Phase 6 (CMIP6) climate data. Future peak flows with 100-year and 500-year return periods are subsequently determined. The latter employs the one-dimensional HEC-RAS model to generate floodplain maps corresponding to 100-year and 500-year flood events. Next, four hazard categories (low, moderate, high, and severe) are produced for each extreme (100-year and 500-year) event. Different risk scores are also assigned to different land use types for vulnerability assessment. Lastly, the risk-maps for both existing and future extreme events are developed, and the risk assessment is conducted by comparing the present risk to the projected risk in the future. The framework was applied to a 32-km long reach of the Neuse River in North Carolina, United States. The results indicated an increase in flood inundation area and thus higher flood hazards and risks in the future in the study area. The authors claimed that the findings of the study have the potential to inform policymakers in terms of preparing flood risk mitigation plans.

Lee et al. [4] developed a Mean Field Bias (MFB)-aware variational (VAR) assimilation framework, or MVAR, to account for catchment-wide biases in distributed hydrologic model states. MVAR corrects the MFB in model states first, and then updates the resulting states at the grid box scale. MVAR was comparatively evaluated with the conventional VAR based on results from assimilating streamflow into the distributed Sacramento Soil Moisture Accounting model for the 2258 km^2 headwater basin draining into the Elk River near Tiff City, Missouri. Compared to the conventional VAR, MVAR reduced the mean squared error of streamflow by 2–8% at the outlet, and by 1–10% at the interior location. MVAR adjusts model states at remote cells from the outlet by larger margins than the VAR.

Giannaros et al. [5] introduced a novel WRF-Hydro-based fluvial flood forecasting system for the Southeast Mediterranean (SEM) and presented modeling methodologies and forecasting schemes used in the system. The initial prototype results derived under a pre-operational mode were presented, and future developments and challenges were discussed.

In conclusion, this Special Issue presented the latest studies on methodological and operational advances in hydrological forecasting.

Funding: This research received no external funding

Institutional Review Board Statement: Not applicable.

Informed Consent Statement: Not applicable.

Data Availability Statement: Not applicable.

Acknowledgments: We sincerely thank the authors who contributed to this Special Issue, as well as the reviewers who provided insightful comments to the authors.

Conflicts of Interest: The authors declare no conflict of interest.

References

1. Wu, L. Tuning the Bivariate Meta-Gaussian Distribution Conditionally in Quantifying Precipitation Prediction Uncertainty. *Forecasting* **2020**, *2*, 1–19. [CrossRef]
2. Bhuiyan, M.A.E.; Yang, F.; Biswas, N.K.; Rahat, S.H.; Neelam, T.J. Machine Learning-Based Error Modeling to Improve GPM IMERG Precipitation Product over the Brahmaputra River Basin. *Forecasting* **2020**, *2*, 248–266. [CrossRef]
3. Pokhrel, I.; Kalra, A.; Rahaman, M.; Thakali, R. Forecasting of Future Flooding and Risk Assessment under CMIP6 Climate Projection in Neuse River, North Carolina. *Forecasting* **2020**, *2*, 323–346. [CrossRef]
4. Lee, H.; Shen, H.; Seo, D.-J. Mean Field Bias-Aware State Updating via Variational Assimilation of Streamflow into Distributed Hydrologic Models. *Forecasting* **2020**, *2*, 526–548. [CrossRef]
5. Giannaros, C.; Galanaki, E.; Kotroni, V.; Lagouvardos, K.; Oikonomou, C.; Haralambous, H.; Giannaros, T. Pre-Operational Application of a WRF-Hydro-Based Fluvial Flood Forecasting System in the Southeast Mediterranean. *Forecasting* **2021**, *3*, 437–446. [CrossRef]

forecasting

MDPI

Communication

Pre-Operational Application of a WRF-Hydro-Based Fluvial Flood Forecasting System in the Southeast Mediterranean

Christos Giannaros [1,2,*], Elissavet Galanaki [2], Vassiliki Kotroni [2], Konstantinos Lagouvardos [2], Christina Oikonomou [1], Haris Haralambous [1,3] and Theodore M. Giannaros [2]

1 Frederick Research Center, Pallouriotissa, Nicosia 1036, Cyprus; res.ec@frederick.ac.cy (C.O.); eng.hh@frederick.ac.cy (H.H.)
2 National Observatory of Athens, Institute for Environmental Research and Sustainable Development, Palea Penteli, 15236 Athens, Greece; galanaki@noa.gr (E.G.); kotroni@noa.gr (V.K.); lagouvar@noa.gr (K.L.); thgian@noa.gr (T.M.G.)
3 Department of Electrical Engineering, Frederick University, Pallouriotissa, Nicosia 1036, Cyprus
* Correspondence: res.gc@frederick.ac.cy

Abstract: The Southeast Mediterranean (SEM) is characterized by increased vulnerability to river/stream flooding. However, impact-oriented, operational fluvial flood forecasting is far away from maturity in the region. The current paper presents the first attempt at introducing an operational impact-based warning system in the area, which is founded on the coupling of a state-of-the-art numerical weather prediction model with an advanced spatially-explicit hydrological model. The system's modeling methodology and forecasting scheme are presented, as well as prototype results, which were derived under a pre-operational mode. Future developments and challenges needed to be addressed in terms of validating the system and increasing its efficiency are also discussed. This communication highlights that standard approaches used in operational weather forecasting in the SEM for providing flood-related information and alerts can, and should, be replaced by advanced coupled hydrometeorological systems, which can be implemented without a significant cost on the operational character of the provided services. This is of great importance in establishing effective early warning services for fluvial flooding in the region.

Keywords: flood; forecasting; hydrometeorology; WRF-Hydro; impact-based; warnings; Cyprus; Greece; Mediterranean

Citation: Giannaros, C.; Galanaki, E.; Kotroni, V.; Lagouvardos, K.; Oikonomou, C.; Haralambous, H.; Giannaros, T.M. Pre-Operational Application of a WRF-Hydro-Based Fluvial Flood Forecasting System in the Southeast Mediterranean. *Forecasting* 2021, 3, 437–446. https://doi.org/10.3390/forecast3020026

Academic Editors: Minxue He, Haksu Lee and Sungwook Wi

Received: 26 April 2021
Accepted: 8 June 2021
Published: 9 June 2021

Publisher's Note: MDPI stays neutral with regard to jurisdictional claims in published maps and institutional affiliations.

1. Introduction

The geomorphological complexity of the Southeast Mediterranean (SEM) has a significant influence on the atmospheric processes in the area, especially on those related to rainfall [1]. The combined effects of large-scale atmospheric circulation, land-sea temperature contrast and orographic forcing produce frequently huge amounts of rainfall, exceeding 100–150 mm even within a few hours and/or with hourly intensities higher than 50 mm [2–6]. Such heavy rainfall episodes favor saturation and/or infiltration excess, and subsequently the occurrence of floods, which can be devastating in terms of damages and human losses, in the numerous small- and medium-sized catchments present in the SEM. Indeed, Vinet et al. [7] found an increasing flood mortality trend in the SEM. This fact highlights the urgent need to provide timely and accurate flood-related information, including warnings and recommendations.

Considering the above, rainfall is traditionally used as an indicator in operational weather forecasting activities in the SEM, for assessing the potential of water-associated disasters' occurrence. However, the onset of a flood incident depends on the hydrological features and processes in the catchment where rain occurs. Thus, combining numerical weather prediction (NWP) and hydrological modeling is required in order to provide a comprehensive assessment of forthcoming atmospheric phenomena and their projected impact on the hydrological response at watershed level. In this direction, significant progress has been made in past decades concerning the development of physically-based and spatially-explicit hydrological models, and their coupling to advanced NWP systems [8,9]. Although the coupling hydrometeorological framework has been adopted in several retrospective forecasting experiments in the SEM [10–15], little documentation exists with respect to operational forecasting practices and concerns (e.g., Israeli Hydrological Service [16]).

The present paper discusses the pre-operational experience in a flood forecasting system targeted over three watersheds in Cyprus (Larnaca region) and Greece (Attica region) in the framework of the first attempt at introducing impact-based flood warnings in the SEM. This experience includes prototype results derived during the tentative application of the system, which started in September 2020. The system is based on the coupling between the Weather Research and Forecasting (WRF) NWP model and its hydrological extension package (WRF-Hydro). The modeling architecture of the system is presented in detail, as well as future advancements with respect to evaluating thoroughly the system's performance and enhancing its efficiency. The key aim of the system is to estimate the potential impact of fluvial flooding in the catchments of interest, and provide relevant information and advice. The presented work is the first of its kind in both Cyprus and Greece. It is also the first step towards establishing an advanced early warning service, capable of interpreting and evaluating hydrometeorological information in terms of flood risk levels and possible effects under an operational context in both countries.

2. Methodology

The flood forecasting system is built upon the state-of-the-art WRF and WRF-Hydro models [17,18]. As illustrated in Figure 1, a special procedure is applied on a daily basis for assessing the necessity of initiating the hydrometeorological forecasting over the targeted catchments. At a first stage, WRF operates once daily (1200 UTC forecasting cycle) to simulate the meteorological conditions for the next 3 days (72 h) over Cyprus and Greece. The second step in the flood forecasting modeling chain involves the examination of the precipitation forecasts, provided over Cyprus and Greece, in the targeted watersheds. In case of high-impact rainfall thresholds exceedance, the WRF-Hydro model is activated under its fully coupled mode (i.e., two-way interaction between the atmospheric and hydrological processes) to provide 36 h streamflow forecasts over the catchments of interest. The last step in the forecasting process involves the post-processing and visualization of the WRF-Hydro output, including flood warnings in case of exceedance of flood-associated streamflow thresholds (Figure 1). It is important to note that the operational modeling methodology demonstrated above allows for scalability (e.g., expansion of the hydrometeorological forecasting over more catchments in Cyprus and Greece). Additionally, it is flexible with respect to computational demands and timeliness of the forecasts' delivery. Concerning the latter, the flood forecasting system is initialized once per day in the late afternoon (1930 UTC) and, in case of its full implementation (i.e., hydrometeorological prediction activation over all three targeted watersheds), the forecasting products are available early in the morning of the next day (~0600 UTC). In particular, the WRF meteorological forecast implementation requires 1.5 h and 3 h over Cyprus and Greece, respectively. The WRF-Hydro simulation for each catchment of interest is executed in 45 min, while post-processing the outcomes of each WRF-Hydro run takes 15 min (Figure 1). Considering that the WRF-Hydro forecasts are extended to a horizon of 36 h, the nighttime period of the model's implementation is overlapped by the previous forecast.

The WRF model configuration is based on the integrated HERMES weather forecasting system developed at the National Observatory of Athens (NOA; [5]). It consists primarily of two two-way nested modeling domains (Figure 2a): (a) d01 covering most of Europe and North Africa to capture synoptic-scale atmospheric conditions (10 km horizontal grid increment), and (b) d02_GR focusing over Greece and neighboring countries (2 km horizontal grid increment). The model uses initial and boundary conditions based on the 0.25° × 0.25° spatial resolution and 6-h temporal resolution Global Forecast System (GFS) operational data. After the weather forecast implementation over Greece, a one-way nesting approach is applied for the 3-day weather forecast over the domain d02_CY that focuses on the greater area of Cyprus (2 km horizontal grid increment; Figure 2a), using the "ndown" WRF capability [17]. The initial and boundary conditions for this finer-grid run are obtained from the coarse domain (d01), while high-resolution terrestrial data are used for initializing the run, as in the case of Greece (d02_GR).

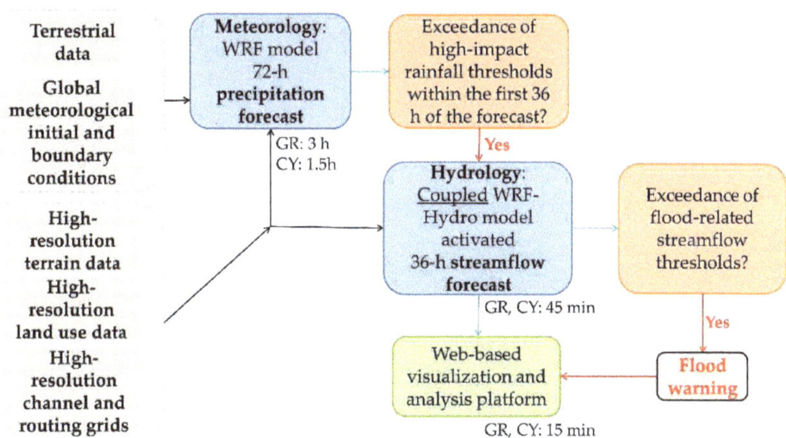

Figure 1. Flowchart of the flood forecasting service. Timings correspond to the implementation of the WRF meteorological and WRF-Hydro hydrological forecasts, and to the post-processing of the forecasts over Cyprus (CY) and Greece (GR).

For examining the precipitation forecasts, provided by the 2 km resolution domains over Cyprus and Greece, in the targeted watersheds, an automated algorithm is applied. The algorithm investigates if any 1 h rainfall or 6 h moving sum of rainfall exceeds 10 mm or 18 mm, respectively, within the first 1.5 days (36 h) of forecast over the Archangelos/Kamitsis river basin in Cyprus (red outline in Figure 2b). For the same forecasting period over Greece, the algorithm examines if any 1 h rainfall or 6 h moving sum of rainfall surpasses 15 mm or 30 mm, respectively, over the Rafina and Sarantapotamos stream basins (red outlines in Figure 2c). The rainfall thresholds for identifying a potential high-impact event were defined based on previous hydrometeorological and impact analysis studies, and experience in forecasting and monitoring weather and natural disasters in the areas [5,19–22].

For the hydrometeorological forecasting, the WRF-Hydro setup includes high-resolution (~667 m horizontal grid increment) domains over Cyprus (d03_CY; Figure 2b) and Attica, Greece (d03_GR; Figure 2c), where the meteorological conditions are simulated. Over these domains, ultra-high-resolution (~95 m horizontal grid increment) domains are configured for simulating the hydrological procedures over the Archangelos/Kamitsis (Figure 2b), Rafina and Sarantapotamos (Figure 2c) watersheds. Under the fully coupled mode, the meteorological forcing to WRF-Hydro is provided by the WRF model. Additionally, specific hydrological-related land surface state variables modeled by WRF-hydro in the ultra-high-resolution domains, including soil moisture, feed back to the WRF-simulated atmospheric processes. This feedback has the potential to enhance the model's rainfall forecast skill, as two recent research works conducted in the SEM have shown [23,24]. The highly-resolved modeling of surface and saturated subsurface water, and the exchange of information between the meteorological and hydrological component of the model, are taking place through a sub-grid aggregation/disaggregation process [18], using ultra-high-resolution terrain, land use and channel/routing grid data. Prior to its pre-operational setup, the WRF-Hydro model was calibrated and validated over the configured domains following a manual stepwise approach based on the comparison of the modeled output with observations [23].

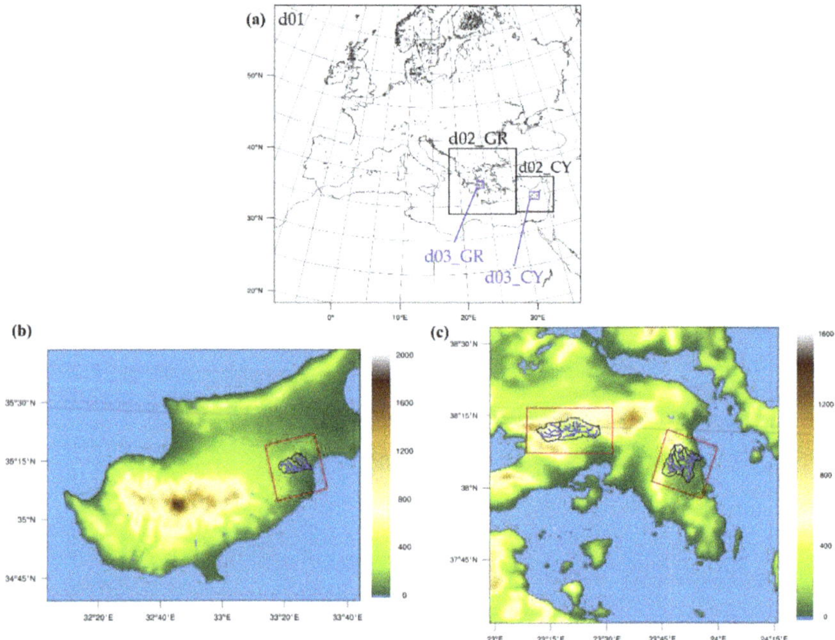

Figure 2. (**a**) WRF modeling domains; (**b**) topography of Cyprus in the d03_CY modeling domain with identification of Archangelos/Kamitsis catchment; (**c**) topography of Attica, Greece in the d03_GR modeling domain with identification of the Rafina (East) and Sarantapotamos (West) stream basins.

During the WRF-Hydro output post-processing, automated computing scripts produce maps that demonstrate the spatial distribution of streamflow in the targeted catchments. Hydrographs illustrating the temporal evolution of river/stream discharge and basin-averaged rainfall are also being created for specific locations at risk. Further, the exceedance of flood-associated streamflow thresholds is being examined at the high-risk locations. Currently, the latter procedure is applied only for the Rafina stream basin, where preliminary impact-based thresholds have been defined according to the outcomes of a recent study on the hydrometeorological and socioeconomic impact assessment in the area [22]. In case the discharge thresholds are exceeded, an alert message is created providing information on the potential impacts in the watershed and on the preparedness level. At present, the products described above are being automatically published in a dedicated web portal (https://w1.meteo.gr/cypjp/index.cfm, accessed on 9 June 2021), which is publicly accessible under a pilot demonstration mode, constructed in the frames of the CyFFORS project (Cyprus Flood Forecasting System, http://cyirg.frederick.ac.cy//cyprus-flood-forecasting-system/, accessed on 9 June 2021).

3. Prototype Results

Figure 2 presents two examples of the gridded streamflow maps that are produced with a temporal resolution of 1 h, when the WRF-Hydro-based hydrometeorological forecasting is activated.

The first map (Figure 3a) was issued for Rafina stream basin (Greece; Figure 2c) on 13 December 2020 and it is valid for 0600 UTC of the next day. The presented case is related to a 3-day period (12–14 December 2020) of heavy rainfall over Attica, Greece, due to the occurrence of two consecutive low-pressure systems that affected the area in that period. On 15 December 2020, the second surface low affected Cyprus, and particularly the Larnaca region, while moving from the west towards the east. Thus, the second map (Figure 3b) refers to the Archangelos/Kamitsis river basin (Cyprus; Figure 2b). The map was issued on 15 December 2020 and it is valid for 1900 UTC of the same day. The interpretation of the results and information shown in the maps is facilitated by (a) the light blue (low streamflow) to red (high streamflow) color map, (b) the identification of regions (written on map) and (c) the background map that highlights the road network. For instance, in Figure 3b, a user can quickly identify the highest values of discharge exceeding 4 m^3/s in the Archangelos/Kamitsis section that passes through the Livadia residential area, close to the river's outlet. Interactive points of interest (i.e., locations at risk) are also provided in the maps (illustrated as yellow circles), so a user can view and even download the 36-h time series of streamflow and basin-averaged rainfall at these points.

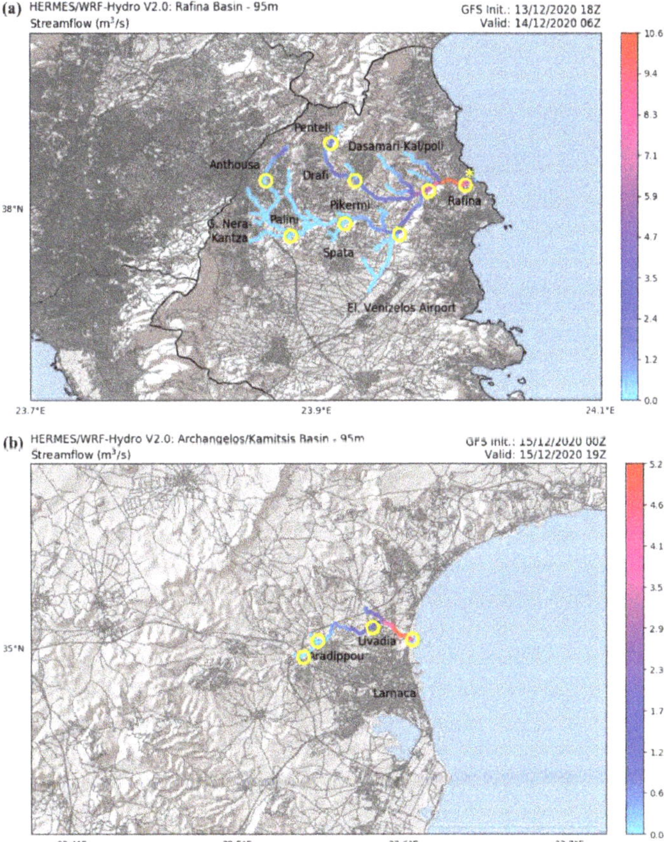

Figure 3. Example of streamflow forecasting map for (**a**) Rafina stream basin (Greece), valid for 14 December 2020 at 0600 UTC, and (**b**) Archangelos/Kamitsis river basin (Cyprus), valid for 15 December 2020 at 1900 UTC. The yellow circles denote specific points of interest.

Figure 4 shows the 36-h time series of streamflow and basin-averaged rainfall at a point of interest targeted over the outlet of the Rafina stream basin (denoted by an asterisk in Figure 3a), as an example of the hydrographs provided in the framework of the hydrometeorological forecasting. In order to communicate flood-related information, which is of great usefulness for both stakeholders and the general public, the hydrographs include the maximum streamflow forecasted within the 1.5-day prediction range at the specific location at risk (Dmax; Figure 4). This information is elaborated in an explanatory description, which is available in the section "HYDRO" of the web portal, where the hydrographs are presented. Further, for the purpose of the impact-based warnings for the Rafina catchment, a table describing the alert level and its corresponding potential impacts is provided (Table 1), following Giannaros et al. [22].

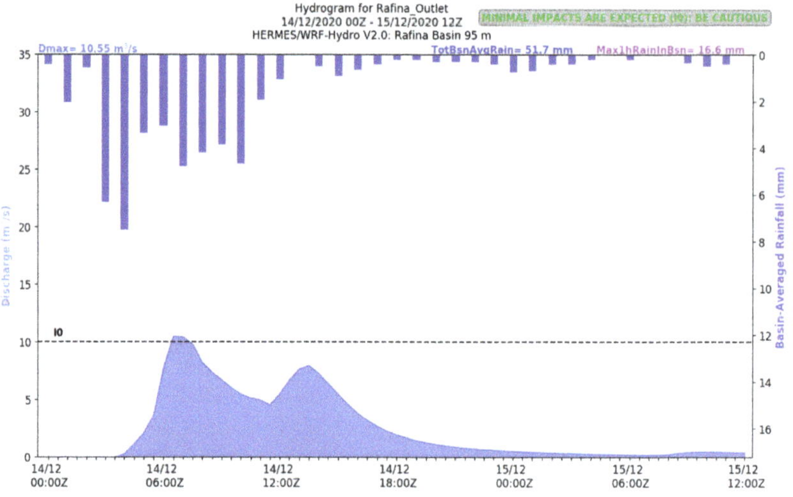

Figure 4. Example of hydrograph forecast for the Rafina stream basin outlet (Attica, Greece), issued on 13 December 2020.

Table 1. Socioeconomic impact intensity classification.

	I0: Minimal	I1: Minor	I2: Significant	I3: Severe
Human life	Not expected	Risk for vulnerable groups of people and/or involved in endangered situations	Danger to life due to physical hazards associated with flooding water	Danger to life due to physical, chemical and utility hazards associated with flooding water
Damage to properties and public structures	Not expected	Light damage to individual properties	Important damage to many properties and/or public structures	Extensive damage to multiple properties and/or public structures
Transportation	Little or no disruption to river crossings and/or roads close to the river/stream	Small-scale disruption (local and short term)	Large-scale disruption (broad and long term) and/or important damage to the transport network	Extensive disruption (broad and long term) and/or extensive damage to the transport network
Utilities	Not expected	Small-scale disruption (local and short term)	Large-scale disruption (broad and long term)	Extensive disruption (broad and long term) and/or loss of utilities

For the provided example over the Rafina stream basin, an alert associated with potential minimal impacts (I0 class; Table 1) was issued, as the maximum streamflow, which was found in the watershed's outlet, exceeded 10 m^3/s (Figure 4). This warning corresponds well to the localized flooding reported in the area by the local media (https://www.irafina.gr/rafina-afti-ine-i-katastasi-stis-gefires-tis-arionos-ke-touvarda-pou-plimmirisan-apo-tin-kakokeria-foto/, accessed on 9 June 2021). On the event day (14 December), 60.2 mm of rain was recorded in the Rafina meteorological station (close to the point of interest denoted by the asterisk in Figure 3a), which is part of the NOA's dense network of automated weather stations [21]. The 24 h rainfall simulated by the WRF-Hydro model at the nearest to the location of the station grid point during the episode was equal to 45.1 mm, indicating an adequate model performance in capturing the observed rainfall's intensity. The encouraging performance of the model is also supported by the observed and modeled rainfall time series (Figure 5). As can be seen in Figure 5, the WRF-Hydro-simulated rainfall was close to the observed one in terms of both timing and intensity.

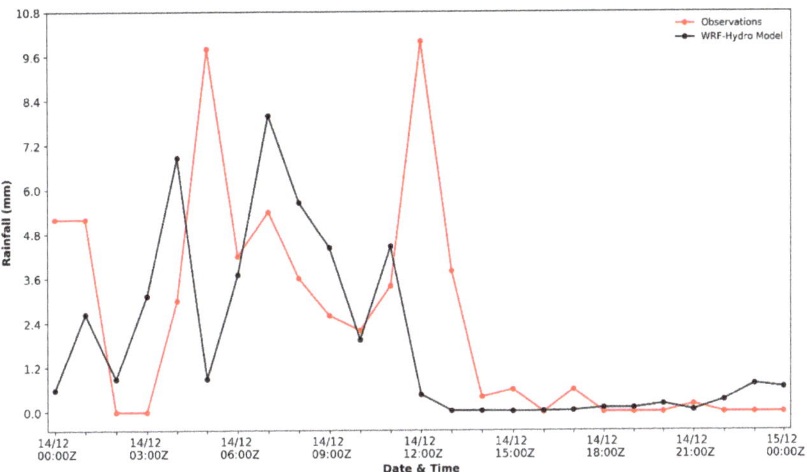

Figure 5. Time series of the observed and WRF-Hydro-modeled rainfall over the Rafina meteorological station from 14 December 2020, 0000 UTC, to 15 December 2020, 0000 UTC.

4. Discussion

Decision-making on the assessment of flood risk and on the issuing of river- and stream-scale alerts is a significant challenge in the context of operational weather forecasting. Combining NWP and hydrological models has been proven to be valuable in addressing this difficult task [25–28]. However, the production and dissemination of flood-related information using advanced hydrometeorological models and convenient communication methods on an operational basis have received little attention in the Southeast Mediterranean.

The current paper presents the modeling framework and pre-operational use of a river/stream flood forecasting system in the SEM region. The system is targeted over three catchments in two highly flood-prone regions (Larnaca in Cyprus and Attica in Greece; Figure 2b,c). It is based on the advanced coupling between the WRF NWP model and the WRF-Hydro hydrological model, exploiting a flexible forecasting scheme in order to provide streamflow predictions when a heavy rainfall event is expected. The river/stream discharge forecasts are compared to specific thresholds in one of the watersheds of interest, namely in the Rafina stream basin, and a warning can be issued depending on the impact that is associated with the potentially reached threshold [22]. Although a comprehensive evaluation of the system's performance has not yet been carried out, the preliminary results indicate that the system has the capability to adequately forecast flood-associated levels of river/stream discharge, as in the case related to localized flooding in the Rafina catchment (Figure 4).

A thorough evaluation of the system in terms of rainfall and streamflow forecasting accuracy is expected to take place after the end of its pre-operational application. Further, potential developments of the system will be examined in the future in order to increase its effectiveness in providing impact-based fluvial flood warnings. These developments include the continuous assessment and improvement of the rainfall forecasts provided by the WRF-based HERMES weather forecasting system (e.g., [5]). Additionally, they could include the introduction of more frequent daily operational runs, instead of the current once-a-day implementation of the system. This approach provides the potential of using the most up-to-date and accurate initialization data. The latter may include real-time rainfall data, exploiting the dense network of automated weather stations operated by NOA [21], as forcing the WRF-Hydro model with observations can enhance the streamflow predictions [24]. In this case, the one-way coupling mode of WRF-Hydro can be utilized.

Concerning the initialization data that are associated with the initial and boundary conditions (IBCs), they are critical to the WRF model's performance, as even a 6-h difference in the IBCs' lead time can result in significant differences in the simulated meteorological fields, including those of rainfall [29]. In the direction of considering the uncertainties in atmospheric IBCs, ensemble forecasting approaches can be beneficial [30], leading to probabilistic streamflow forecasts that can assist in quantifying the likelihood of impact-based thresholds' exceedance [31]. However, it should be noted that IBCs' ensemble forecasting introduces increased demands for computing resources and thus, it may not be feasible under an operational context. An alternative approach could be a multi-model ensemble using meteorological forcing data derived from different NWP models, for the one-way coupled WRF-Hydro model. Beyond the WRF model, these could include the BOLAM [32], MOLOCH, ICON and APREGE models, which are currently used by NOA for its operational weather forecasting activities. Finally, future work should focus on defining impact-based discharge thresholds in the Archangelos/Kamitsis river basin and Sarantapotamos catchment, as soon as data from flood events are available, as well as on the expansion of the forecasting system over more watersheds in Cyprus and Greece.

Author Contributions: Conceptualization, C.G., E.G., V.K., K.L., C.O. and H.H; methodology, C.G., E.G., V.K., K.L.; software, C.G, E.G. and T.M.G.; validation, C.G..; formal analysis, C.G.; investigation, C.G.; resources, V.K and K.L..; data curation, C.G.; writing—original draft preparation, C.G.; writing—review and editing, C.G., E.G., V.K., K.L., C.O., H.H. and T.M.G.; visualization, C.G.; supervision, V.K., K.L., C.O. and H.H; project administration, C.O.; funding acquisition, C.G., V.K., K.L., C.O. and H.H. All authors have read and agreed to the published version of the manuscript.

Funding: This research was funded by the project Cyprus Flood Forecasting System—POST-DOC/0718/0040 which is co-funded by the Republic of Cyprus and the European Regional Development Fund (through the 'DIDAKTOR' RESTART 2016–2020 Program for Research, Technological Development and Innovation).

Institutional Review Board Statement: Not applicable.

Informed Consent Statement: Not applicable.

Data Availability Statement: The data presented in this study are available on request from the corresponding author.

Conflicts of Interest: The authors declare no conflict of interest.

References

1. Michaelides, S.; Karacostas, T.; Sánchez, J.L.; Retalis, A.; Pytharoulis, I.; Homar, V.; Romero, R.; Zanis, P.; Giannakopoulos, C.; Bühl, J.; et al. Reviews and perspectives of high impact atmospheric processes in the Mediterranean. *Atmos. Res.* **2018**, *208*, 4–44. [CrossRef]

2. Kotroni, V.; Lagouvardos, K.; Defer, E.; Dietrich, S.; Porcù, F.; Medaglia, C.M.; Demirtas, M. The Antalya 5 December 2002 Storm: Observations and Model Analysis. *J. Appl. Meteorol. Climatol.* **2006**, *45*, 576–590. [CrossRef]

3. Nicolaides, K.A.; Michaelides, S.C.; Savvidou, K.; Orphanou, A.; Constantinides, P.; Charalambous, M.; Michaelides, M. Case studies of selected Project "flash" events. *Adv. Geosci.* **2008**, *17*, 93–98. [CrossRef]

4. Tolika, K.; Maheras, P.; Anagnostopoulou, C. The exceptionally wet year of 2014 over Greece: A statistical and synoptical-atmospheric analysis over the region of Thessaloniki. *Theor. Appl. Climatol.* **2018**, *132*, 809–821. [CrossRef]

5. Giannaros, C.; Kotroni, V.; Lagouvardos, K.; Giannaros, M.T.; Pikridas, C. Assessing the Impact of GNSS ZTD Data Assimilation into the WRF Modeling System during High-Impact Rainfall Events over Greece. *Remote Sens.* **2020**, *12*, 383. [CrossRef]

6.   Lagouvardos, K.; Dafis, S.; Giannaros, C.; Karagiannidis, A.; Kotroni, V. Investigating the role of extreme synoptic patterns and complex topography during two heavy rainfall events in Crete in February 2019. *Climate* **2020**, *8*, 87. [CrossRef]
7.   Vinet, F.; Bigot, V.; Petrucci, O.; Papagiannaki, K.; Llasat, M.C.; Kotroni, V.; Boissier, L.; Aceto, L.; Grimalt, M.; Llasat-Botija, M.; et al. Mapping flood-related mortality in the mediterranean basin. Results from the MEFF v2.0 DB. *Water* **2019**, *11*, 2196. [CrossRef]
8.   Pagano, T.C.; Wood, A.W.; Ramos, M.-H.; Cloke, H.L.; Pappenberger, F.; Clark, M.P.; Cranston, M.; Kavetski, D.; Mathevet, T.; Sorooshian, S.; et al. Challenges of Operational River Forecasting. *J. Hydrometeorol.* **2014**, *15*, 1692–1707. [CrossRef]
9.   Ning, L.; Zhan, C.; Luo, Y.; Wang, Y.; Liu, L. A review of fully coupled atmosphere-hydrology simulations. *J. Geogr. Sci.* **2019**, *29*, 465–479. [CrossRef]
10.   Yucel, I.; Onen, A.; Yilmaz, K.K.; Gochis, D.J. Calibration and evaluation of a flood forecasting system: Utility of numerical weather prediction model, data assimilation and satellite-based rainfall. *J. Hydrol.* **2015**, *523*, 49–66. [CrossRef]
11.   Varlas, G.; Anagnostou, M.N.; Spyrou, C.; Papadopoulos, A.; Kalogiros, J.; Mentzafou, A.; Michaelides, S.; Baltas, E.; Karymbalis, E.; Katsafados, P. A multi-platform hydrometeorological analysis of the flash flood event of 15 November 2017 in Attica, Greece. *Remote Sens.* **2019**, *11*, 45. [CrossRef]
12.   Papaioannou, G.; Varlas, G.; Terti, G.; Papadopoulos, A.; Loukas, A.; Panagopoulos, Y.; Dimitriou, E. Flood inundation mapping at ungauged basins using coupled hydrometeorological-hydraulic modelling: The catastrophic case of the 2006 Flash Flood in Volos City, Greece. *Water* **2019**, *11*, 2328. [CrossRef]
13.   Spyrou, C.; Varlas, G.; Pappa, A.; Mentzafou, A.; Katsafados, P.; Papadopoulos, A.; Anagnostou, M.N.; Kalogiros, J. Implementation of a Nowcasting Hydrometeorological System for Studying Flash Flood Events: The Case of Mandra, Greece. *Remote Sens.* **2020**, *12*, 2784. [CrossRef]
14.   Camera, C.; Bruggeman, A.; Zittis, G.; Sofokleous, I.; Arnault, J. Simulation of extreme rainfall and streamflow events in small Mediterranean watersheds with a one-way-coupled atmospheric-hydrologic modelling system. *Nat. Hazards Earth Syst. Sci.* **2020**, *20*, 2791–2810. [CrossRef]
15.   Ozkaya, A.; Akyurek, Z. WRF-Hydro Model Application in a Data-Scarce, Small and Topographically Steep Catchment in Samsun, Turkey. *Arab. J. Sci. Eng.* **2020**, *45*, 3781–3798. [CrossRef]
16.   Givati, A.; Fredj, E.; Silver, M. Chapter 6: Operational Flood Forecasting in Israel. In *Flood Forecast.*; Adams, T.E., Pagano, T., Eds.; Academic Press: Boston, MA, USA, 2016; pp. 153–167. ISBN 978-0-12-801884-2.
17.   Skamarock, W.C.; Klemp, J.B.; Dudhia, J.; Gill, D.O.; Liu, Z.; Berner, J.; Wang, W.; Powers, J.G.; Duda, M.G.; Barker, D.; et al. *A Description of the Advanced Research WRF Model Version 4*; NCAR: Boulder, CO, USA, 2019.
18.   Gochis, D.J.; Barlage, M.; Dugger, A.; FitzGerald, K.; Karsten, L.; McAllister, M.; McCreight, J.; Mills, J.; Rafieei Nasab, A.; Read, L.; et al. *The WRF-Hydro Modeling System Technical Description, (Version 3.0)*; NCAR Technical Note: Boulder, CO, USA, 2015.
19.   Papagiannaki, K.; Lagouvardos, K.; Kotroni, V. A database of high-impact weather events in Greece: A descriptive impact analysis for the period 2001–2011. *Nat. Hazards Earth Syst. Sci.* **2013**, *13*, 727–736. [CrossRef]
20.   Papagiannaki, K.; Lagouvardos, K.; Kotroni, V.; Bezes, A. Flash flood occurrence and relation to the rainfall hazard in a highly urbanized area. *Nat. Hazards Earth Syst. Sci.* **2015**, *15*, 1859–1871. [CrossRef]
21.   Lagouvardos, K.; Kotroni, V.; Bezes, A.; Koletsis, I.; Kopania, T.; Lykoudis, S.; Mazarakis, N.; Papagiannaki, K.; Vougioukas, S. The automatic weather stations NOANN network of the National Observatory of Athens: Operation and database. *Geosci. Data J.* **2017**, *4*, 4–16. [CrossRef]
22.   Giannaros, C.; Kotroni, V.; Lagouvardos, K.; Oikonomou, C.; Haralambous, H.; Papagiannaki, K. Hydrometeorological and Socio-Economic Impact Assessment of Stream Flooding in Southeast Mediterranean: The Case of Rafina Catchment (Attica, Greece). *Water* **2020**, *12*, 2426. [CrossRef]
23.   Galanaki, E.; Lagouvardos, K.; Kotroni, V.; Giannaros, T.; Giannaros, C. Implementation of WRF-Hydro at two drainage basins in the region of Attica, Greece. *Nat. Hazards Earth Syst. Sci. Discuss.* **2020**, *2020*, 1–28. (In production)
24.   Givati, A.; Gochis, D.; Rummler, T.; Kunstmann, H. Comparing One-Way and Two-Way Coupled Hydrometeorological Forecasting Systems for Flood Forecasting in the Mediterranean Region. *Hydrology* **2016**, *3*, 19. [CrossRef]
25.   Price, D.; Hudson, K.; Boyce, G.; Schellekens, J.; Moore, R.J.; Clark, P.; Harrison, T.; Connolly, E.; Pilling, C. Operational use of a grid-based model for flood forecasting. *Proc. Inst. Civ. Eng. Water Manag.* **2012**, *165*, 65–77. [CrossRef]
26.   Cranston, M.D.; Tavendale, A.C.W. Advances in operational flood forecasting in Scotland. *Proc. Inst. Civ. Eng. Water Manag.* **2012**, *165*, 79–87. [CrossRef]
27   Cohen, S.; Praskievicz, S.; Maidment, D.R. Featured Collection Introduction: National Water Model. *JAWRA J. Am. Water Resour. Assoc.* **2018**, *54*, 767–769. [CrossRef]
28.   Lahmers, T.M.; Gupta, H.; Castro, C.L.; Gochis, D.J.; Yates, D.; Dugger, A.; Goodrich, D.; Hazenberg, P. Enhancing the Structure of the WRF-Hydro Hydrologic Model for Semiarid Environments. *J. Hydrometeorol.* **2019**, *20*, 691–714. [CrossRef]
29.   Avolio, E.; Cavalcanti, O.; Furnari, L.; Senatore, A.; Mendicino, G. Brief communication: Preliminary hydro-meteorological analysis of the flash flood of 20 August 2018 in Raganello Gorge, southern Italy. *Nat. Hazards Earth Syst. Sci.* **2019**, *19*, 1619–1627. [CrossRef]
30.   Furnari, L.; Mendicino, G.; Senatore, A. Hydrometeorological ensemble forecast of a highly localized convective event in the mediterranean. *Water* **2020**, *12*, 1545. [CrossRef]

31. Maxey, R.; Cranston, M.; Tavendale, A.; Buchanan, P. The use of deterministic and probabilistic forecasting in countrywide flood guidance in Scotland. In *Proceedings* of the BHS Eleventh National Symposium, Hydrology for a Changing World, Dundee, UK, 9–11 July 2012.
32. Lagouvardos, K.; Kotroni, V.; Koussis, A.; Feidas, H.; Buzzi, A.; Malguzzi, P. The Meteorological Model BOLAM at the National Observatory of Athens. *J. Appl. Meteorol.* **2003**, *42*, 1667–1678. [CrossRef]

Article

Mean Field Bias-Aware State Updating via Variational Assimilation of Streamflow into Distributed Hydrologic Models

Haksu Lee [1,*], Haojing Shen [2] and Dong-Jun Seo [2]

[1] Len Technologies, Oak Hill, VA 20171, USA

[2] Department of Civil Engineering, University of Texas at Arlington, Arlington, TX 76019, USA; haojing.shen@mavs.uta.edu (H.S.); djseo@uta.edu (D.-J.S.)

*   Correspondence: haksu.lee@noaa.gov; Tel.: +(1)240-413-1133

Received: 1 November 2020; Accepted: 10 December 2020; Published: 11 December 2020

Abstract: When there exist catchment-wide biases in the distributed hydrologic model states, state updating based on streamflow assimilation at the catchment outlet tends to over- and under-adjust model states close to and away from the outlet, respectively. This is due to the greater sensitivity of the simulated outlet flow to the model states that are located more closely to the outlet in the hydraulic sense, and the subsequent overcompensation of the states in the more influential grid boxes to make up for the larger scale bias. In this work, we describe Mean Field Bias (MFB)-aware variational (VAR) assimilation, or MVAR, to address the above. MVAR performs bi-scale state updating of the distributed hydrologic model using streamflow observations in which MFB in the model states are first corrected at the catchment scale before the resulting states are adjusted at the grid box scale. We comparatively evaluate MVAR with conventional VAR based on streamflow assimilation into the distributed Sacramento Soil Moisture Accounting model for a headwater catchment. Compared to VAR, MVAR adjusts model states at remote cells by larger margins and reduces the Mean Squared Error of streamflow analysis by 2–8% at the outlet Tiff City, and by 1–10% at the interior location Lanagan.

Keywords: mean field bias; data assimilation; distributed hydrologic model; streamflow

1. Introduction

Streamflow observations are used extensively to update hydrologic model states via various forms of data assimilation (DA) [1–12]. Assimilating streamflow data into distributed models compared to that into lumped models presents additional challenges due to the greatly increased dimensionality of the inverse problem as elaborated below. In distributed modeling, changes in model states at a cell distant from a high-order stream exert a much smaller influence on the model-simulated flow at the catchment outlet compared to those at a cell near a high-order stream. This is because the runoff generated near a low-order stream, that is, in a headwater area, has to travel much longer distances to reach a high-order stream where channel flow takes over. Most of the above travel occurs on hillslopes where the flow is greatly attenuated. Unlike channel flow, hillslope flow occurs through numerous flow paths on the land surface and pore spaces in the subsurface. These flow paths vary greatly in size, length and shape, and have much larger friction factors than channels. Compared to channel flow, hillslope flow is hence subject to mechanical dispersion and is characterized by smaller ratios of advective to diffusive transport [13]. Recall in the advection-diffusion solution [14,15] that the width of a pulse inflow varies approximately with $\sqrt{Dt}$ where D and t denote the diffusivity and time elapsed, respectively. Larger D and t due to dispersion and longer travel time mean that hillslope flow is greatly attenuated and that sensitivity of changes in runoff generation to lateral inflow into channels via hillslope routing is dampened out for DA to effectively source-trace. Consequently, the sensitivity of the model-simulated streamflow at the outlet to the changes in the model states is smoothed out in the upstream areas of the catchment. Most data assimilation (DA) methods use this spatiotemporal pattern of sensitivity for gradient-based minimization at the grid box scale whether it is evaluated via adjoint code as in variational assimilation (VAR, [16]) or estimated via random sampling in ensemble subspace as in ensemble Kalman filter (EnKF, [17]). The above observation indicates that, when there exist catchment-wide biases, DA at the grid box scale is likely to over- and under-adjust the model states close to and distant from high-order streams to compensate for the large-scale bias. The premise of this work is that one may improve the performance of DA significantly by updating the model states at the catchment scale first and then the resulting states at the grid box scale.

Correcting spatiotemporal biases such as MFB, local bias (LB) and conditional bias (CB) in the data or modeled variables has been extensively explored in statistical pre- and post-processing, calibration, and DA. Bias correction has also been shown to be effective in improving estimation and prediction of soil moisture [18,19], precipitation [20,21], streamflow [22] and extremes [23] as well as forecast models [24], and state updating [25]. In lumped hydrologic modeling, streamflow prediction is improved by correcting biases in forcing forecasts [26,27] or by accounting for CBs in model soil moisture [9]. Correcting MFB has been widely used in gridded precipitation analysis [28–30], which in turn improves streamflow prediction [11]. In distributed hydrologic modeling, correcting MFB in radar-based precipitation estimates improved streamflow and soil moisture [11]. In calibration, adjusting model parameters spatially uniformly also helps preserve their spatial structure [31]. While diverse bias correction methods have emerged in hydrology, correcting the MFB in states of a distributed hydrologic model is new and warrants the development of the MFB-aware DA framework and the assessment of its practicality.

In this work, we describe and evaluate MFB-aware VAR, referred to herein as MVAR, for state updating of distributed hydrologic models based on variational assimilation of streamflow data. MVAR adjusts the model states at the catchment scale first and adjusts the resulting states at the grid box scale. One may hence consider MVAR the simplest form of multiscale DA [32,33] employing two scales, that is, the entire catchment and a grid box. Catchment-wide adjustment in the first step mitigates over- and under-correction of model states at the grid box scale and helps preserve the spatial structure of the background states in the updated fields. Lee et al. [8] have shown that basin-wide uniform adjustment of model states is often able to produce flow predictions that are as accurate as those from adjusting states at individual cells. In such cases, the first step of MVAR would be the main contributor to DA performance and also limit the changes to the background states to a minimum. Note that, if two DA solutions have the same predictive skill, the one with smaller adjustments to the background states should be preferred [33,34]. The effectiveness of MVAR depends also on the goodness of the hydrologic models. If significant parametric or structural errors exist, they are likely to compromise the performance of MVAR. To assess the impact of model errors, we carry out an evaluation using both weakly- and strongly-constrained DA, the latter of which does not assume model errors. The significant new contributions of this paper are the development of MFB-aware DA for streamflow assimilation into the distributed model and the comparative analysis and evaluation of MVAR and VAR. This paper is organized as follows. Section 2 describes the hydrologic model used and the two assimilation approaches, that is, MVAR and VAR. Section 3 describes the study basin and evaluation metrics used. Sections 4 and 5 present results and conclusions, respectively.

2. Model and Data Assimilation

Since the formulation of the assimilation problem is specific to the model used, we first describe the hydrologic model in Section 2.1, followed by formulating the problem in the case of MVAR and VAR in Sections 2.2 and 2.3, respectively.

2.1. Hydrological Model

The model used is the distributed Sacramento Soil Moisture Accounting (SAC-SMA; [35]) included in the US National Weather Service Hydrology Laboratory's Research Distributed Hydrologic Model [31]. The distributed SAC-SMA model operates on the Hydrologic Rainfall Analysis Project (HRAP; [36]) grid mesh of about 4 km in size where NWS radar and multisensor precipitation estimates are available. For potential evaporation, monthly climatology on the HRAP grid is used [7]. The SAC-SMA computes fast and slow runoff components from two subsurface storages, namely, the Upper Zone (UZ) and the Lower Zone (LZ) where the LZ is typically thicker than the UZ. Table 1 summarizes tension (TWC) and free water contents (FWC) in the UZ and LZ.

Table 1. Model states updated via MVAR and VAR for the distributed SAC-SMA model.

SAC-SMA Model States	Description
UZTWC	Upper Zone Tension Water Content
UZFWC	Upper Zone Free Water Content
LZTWC	Lower Zone Tension Water Content
LZPFC	Lower Zone Primary Free Water Content
LZSFC	Lower Zone Supplemental Free Water Content
ADIMC	Additional impervious area water content

Tension water contents (UZTWC and LZTWC) are related to soil moisture bounded to soil particles and free water contents (UZFWC, LZPFC, LZSFC) are related to soil gravitational water. Primary (LZPFC) and supplemental (LZSFC) free water contents in the LZ produce slow- or fast-responding baseflow, respectively. Tension and free soil moisture contents and the additional impervious area water content (ADIMC) interact with each other and generate baseflow, interflow, surface runoff, direct runoff, and impervious area runoff. The kinematic-wave routing model is then used to route the runoff through the hillslopes and channels based on cell-to-cell connectivity information created from the cell outlet tracing with an area threshold (COTAT) algorithm [31,37]. A priori parameters of the distributed SAC-SMA were estimated from the STATSGO2 soil data [38] and the routing parameters were from the digital elevation model (DEM) and channel hydraulic data [31]. This study used manually-optimized model parameters obtained from Phase 1 of the Distributed Model Intercomparison Project (DMIP, [39]). The soil moisture translation routine of the SAC-SMA with heat transfer dynamics (SAC-HT; [40]) is used to calculate depth-specific soil moisture contents from the original SAC-SMA model states.

2.2. MFB-Aware Variational Data Assimilation, MVAR

The following states the data assimilation problem solved in this study. Given the a priori model states, streamflow observations at either outlet or both outlet and interior locations, and observed precipitation (P) and monthly climatology of potential evapotranspiration (PE), update the distributed model soil moisture states at the beginning of the assimilation window and multiplicative biases in P and PE. The model may be applied as a weak constraint to this problem by considering model inadequacies [41–43].

MVAR first adjusts model states via multiplying MFB estimates and then individually updates at the HRAP scale. With enough length of the assimilation window, the MFB estimate effectively affects interior points through propagating outlet flow information to upstream pixels via the routing process, not just affected by high sensitivity area around the basin outlet pixel. The two-step approach in estimating MFB and HRAP cell-scale errors is to reduce the dimensionality or ill-posedness of the assimilation problem. Similar examples include dual state-parameter estimation [44] or data assimilation by field alignment to solve separately displacements and amplitudes [45]. The MFB estimation step can be formulated as the nonlinear constrained least-squares minimization problem in Equations (1) and (2).

Minimize

$$J_K(\mathbf{X}_B, \mathbf{X}_P, \mathbf{X}_E, \mathbf{X}_W) = \tfrac{1}{2}[\mathbf{Z}_B - \mathbf{H}_B\mathbf{X}_B]^T\mathbf{R}_B^{-1}[\mathbf{Z}_B - \mathbf{H}_B\mathbf{X}_B]$$
$$+\tfrac{1}{2}[\mathbf{Z}_P - \mathbf{H}_P\mathbf{X}_P]^T\mathbf{R}_P^{-1}[\mathbf{Z}_P - \mathbf{H}_P\mathbf{X}_P]$$
$$+\tfrac{1}{2}[\mathbf{Z}_E - \mathbf{H}_E\mathbf{X}_E]^T\mathbf{R}_E^{-1}[\mathbf{Z}_E - \mathbf{H}_E\mathbf{X}_E]$$
$$+\tfrac{1}{2}\big[\mathbf{Z}_Q - \mathbf{H}_Q(\mathbf{X}_{S,K-L}, \mathbf{X}_P, \mathbf{X}_E, \mathbf{X}_W)\big]^T\mathbf{R}_Q^{-1}\big[\mathbf{Z}_Q - \mathbf{H}_Q(\mathbf{X}_{S,K-L}, \mathbf{X}_P, \mathbf{X}_E, \mathbf{X}_W)\big]$$
$$+\tfrac{1}{2}\mathbf{X}_W^T\mathbf{R}_W^{-1}\mathbf{X}_W \tag{1}$$

subject to

$$\begin{cases} \mathbf{X}_{S,k} = \mathbf{M}\big(\mathbf{X}_{S,k-1}, \mathbf{X}_P, \mathbf{X}_E\big), & k = K - L + 1, \ldots, K \\ X_{S,j,i}^{min} \leq X_{S,j,i,k} \leq X_{S,j,i}^{max}, & k = K - L, \ldots, K; j = 1, \ldots, n_S; i = 1, \ldots, n_C \end{cases} \tag{2}$$

In Equation (1), $\mathbf{Z}_B = \mathbf{X}_{S,K-L}$ and $\mathbf{H}_B = \mathbf{X}_{S,K-L}^T$. $\mathbf{X}_B = \big[X_{B,j}\big]^T$ is the vector composed of MFB for individual model states where $j = 1, \ldots, n_S$; J_K denotes the objective function value at hour K; $\mathbf{X}_S$ denotes the state vector of the SAC-SMA model, or UZTWC, UZFWC, LZTWC, LZSFC, LZPFC, and ADIMC (Table 1); $\mathbf{X}_{S,K-L}$ represents $\mathbf{X}_S$ at the beginning ($k = K - L$) of the assimilation window, that is, background model states as driven by the base model simulation; $\mathbf{X}_P$ and $\mathbf{X}_F$ denote the multiplicative adjustment factors for the precipitation and Potential Evaporation (PE) data at $k = K - L + 1, \ldots, K$, respectively, within the assimilation window; the subscript k denotes the time index; $X_{S,j,i}^{min}$ and $X_{S,j,i}^{max}$ denote the lower and upper bounds of the j-th state variable at the i-th grid, $X_{S,j,i}$; n_S denotes the number of SAC states ($n_S = 6$ in this study for both distributed and lumped SAC-SMA); n_C denotes the number of HRAP cells within the basin; L denotes the length of the assimilation window—the duration of the unit hydrograph is used for L which represents the time scale of the fast response of the basin; $\mathbf{M}()$ denotes the SAC-SMA model; $\mathbf{H}_B$, $\mathbf{H}_P$, $\mathbf{H}_E$, and $\mathbf{H}_Q()$ denote the observation operators that relate the control vector ($\mathbf{X}_{S,K-L}, \mathbf{X}_P, \mathbf{X}_E, \mathbf{X}_W$) to observations ($\mathbf{Z}_B, \mathbf{Z}_P, \mathbf{Z}_E, \mathbf{Z}_Q$) where, in this study, $\mathbf{H}_B = \mathbf{I}$, $\mathbf{H}_P = \mathbf{Z}_P$, $\mathbf{H}_E = \mathbf{Z}_E$, and $\mathbf{H}_Q()$ is the distributed SAC-SMA and kinematic-wave routing models; $\mathbf{R}_B$, $\mathbf{R}_P$, $\mathbf{R}_E$, $\mathbf{R}_Q$, and $\mathbf{R}_W$ denote the measurement error covariance matrices for background model states, precipitation, PE, streamflow, and the model error, respectively; $\mathbf{Z}_P$, $\mathbf{Z}_E$, and $\mathbf{Z}_Q$ denote the observations of precipitation, PE, and streamflow at $k = K - L + 1, \ldots, K$, respectively; Equation (1) is based on the following observation equations:

$$\mathbf{Z}_B = \mathbf{H}_B\mathbf{X}_B + \mathbf{V}_B \tag{3}$$

$$\mathbf{Z}_P = \mathbf{H}_P\mathbf{X}_P + \mathbf{V}_P \tag{4}$$

$$\mathbf{Z}_E = \mathbf{H}_E\mathbf{X}_E + \mathbf{V}_E \tag{5}$$

$$\mathbf{Z}_Q = \mathbf{H}_Q(\mathbf{X}_{S,K-L}, \mathbf{X}_P, \mathbf{X}_E, \mathbf{X}_W) + \mathbf{V}_Q \tag{6}$$

$$\mathbf{Z}_W = \mathbf{H}_W(\mathbf{X}_{S,K-L}, \mathbf{X}_P, \mathbf{X}_F, \mathbf{X}_W) + \mathbf{V}_W \tag{7}$$

In the above, $\mathbf{V}_B$, $\mathbf{V}_P$, $\mathbf{V}_E$, $\mathbf{V}_Q$, and $\mathbf{V}_W$ denote measurement error vectors for background model states, precipitation, PE, streamflow, and the rainfall-runoff model error. The $\mathbf{Z}_W$ denotes (unknown) observations of an error in the rainfall-runoff model. Similar to Equation (1) in Beven [46], the rainfall-runoff model error $\mathbf{X}_W$ may relate to runoff observations ($\mathbf{Z}_R$) by Equation (8).

$$\mathbf{Z}_R = \mathbf{H}_R(\mathbf{X}_{S,K-L}, \mathbf{X}_P, \mathbf{X}_E, \mathbf{X}_W) + \mathbf{V}_R = \mathbf{M}_R(\mathbf{X}_{S,K-L}, \mathbf{X}_P, \mathbf{X}_E) + \mathbf{X}_W + \mathbf{V}_R \tag{8}$$

In Equation (8), $\mathbf{V}_R$ denotes the measurement error vector for runoff and $\mathbf{M}_R$ represents the rainfall-runoff model. Note in Equation (8) that, in order to model the error in a rainfall-runoff model, $\mathbf{X}_W$ is added to the model-generated runoff $\mathbf{M}_R$—to reflect this into the model source code, the code can be modified by adding $\mathbf{X}_W$ to the model generated runoff, set aside including the $\mathbf{X}_W$

term in the objective function, Equation (1). Since control variables are the only variables adjusted or updated via DA, $\mathbf{M_R}$ is presented as a function of $\mathbf{X}_{S,K-L}$, $\mathbf{X_P}$, and $\mathbf{X_E}$ in Equation (8). Theoretically, however, $\mathbf{M_R}$ should also be a function of model parameters, space-time resolutions, etc. In this respect, $\mathbf{X_W}$ can be interpreted as an agglomerated error of all unresolved error sources propagated by fluxes through model dynamics to the point of generating runoff. If a systematic model bias exists in reproducing streamflow, this may be reflected in the long-term mean $\mathbf{X_W}$ of a large positive or negative value. Equation (8) may be separately written for slow and fast responding runoff components, or collectively for the total channel inflow (TCI). Other important considerations include how to specify the space-time structure of $\mathbf{X_W}$ and its error statistics. When solving the objective function J in Equation (1), $\mathbf{X_W}$ is initially assumed zero, that is, no error in the rainfall-runoff transformation processes; during the minimization, $\mathbf{X_W}$ is changed based on the gradient of J with respect to $\mathbf{X_W}$ in a way to minimize J. Adding $\mathbf{X_W}$ to the runoff is equivalent to adding uncertainties to the initial conditions of the routing model.

To model separately errors in fast and slow runoff components, $X_{W,k,i}^{SURF}$ and $X_{W,k,i}^{GRND}$ are added to the surface and ground runoff, respectively where $X_{W,k,i}^{SURF}$, and $X_{W,k,i}^{GRND}$ represent the model error in surface and ground runoff at the k-th time step at the i-th HRAP grid, respectively. Since the statistical properties of observation errors in surface and ground runoff are unknown in reality, these observation errors are assumed non-informative which drops $\frac{1}{2}\mathbf{X_W}^T\mathbf{R}_W^{-1}\mathbf{X_W}$ in Equation (1) [34]. Assuming that observational errors are independent of one another and time-invariant, observation error covariance matrices $\mathbf{R}$ in Equation (1) become diagonal and static (See [7,47] for justification). This renders Equation (1) rewritten as Equation (9).

Minimize

$$
\begin{aligned}
J_K(\mathbf{X_B}, \mathbf{X_P}, \mathbf{X_E}, \mathbf{X_W}) = {} & \tfrac{1}{2} \sum_{j=1}^{n_S} \sum_{i=1}^{n_C} Z_{B,j,i,K-L}^2 \big[1 - X_{B,j}\big]^2 \sigma_{B,j}^{-2} \\
& + \tfrac{1}{2} \sum_{k=K-L+1}^{K} \sum_{l=1}^{n_G} \big[Z_{Q,l,k} - H_{Q,l,k}(\mathbf{X}_{S,K-L}, \mathbf{X_P}, \mathbf{X_E}, \mathbf{X_W})\big]^2 \sigma_Q^{-2} \\
& + \tfrac{1}{2} \sum_{k=K-L+1}^{K} \sum_{i=1}^{n_C} Z_{P,i,k}^2 \big[1 - X_{P,k}\big]^2 \sigma_P^{-2} \\
& + \tfrac{1}{2} \sum_{k=K-L+1}^{K} \sum_{i=1}^{n_C} Z_{PE,i,k}^2 \big[1 - X_{E,k}\big]^2 \sigma_E^{-2}
\end{aligned}
\tag{9}
$$

subject to

$$
\begin{cases}
\mathbf{X}_{S,k} = \mathbf{M}\big(\mathbf{X}_{S,k-1}, \mathbf{X_P}, \mathbf{X_E}\big), & k = K-L+1, \ldots, K \\
X_{S,j,i}^{min} \leq X_{S,j,i,k} \leq X_{S,j,i}^{max}, & k = K-L, \ldots, K; j = 1, \ldots, n_S; i = 1, \ldots, n_C
\end{cases}
\tag{10}
$$

From minimizing Equation (9), $\mathbf{X_B} = \big[X_{B,j}\big]^T$ is estimated where $X_{B,j}$ denotes the MFB estimate of the j-th model state. Using the estimated $\mathbf{X_B}$ vector, the model state at the individual HRAP grid is updated by minimizing $J_K(\mathbf{X}_{S,K-L}, \mathbf{X_P}, \mathbf{X_E}, \mathbf{X_W})$, which is the same as Equation (9) except replacing $\frac{1}{2}\sum_{j=1}^{n_S}\sum_{i=1}^{n_C} Z_{B,j,i,K-L}^2 \big[1 - X_{B,j}\big]^2 \sigma_{B,j}^{-2}$ in Equation (9) with $\frac{1}{2}\sum_{j=1}^{n_S}\sum_{i=1}^{n_C} X_{B,j}^2 \big[Z_{B,j,i,K-L} - X_{S,j,i,K-L}\big]^2 \sigma_{B,j}^{-2}$.

In updating model states at the individual HRAP grid, the observation equation used for model states is the following:

$$
\widetilde{\mathbf{Z}}_B = \mathbf{H}_B \widetilde{\mathbf{X}}_{S,K-L} + \mathbf{V}_B
\tag{11}
$$

In Equation (11), $\widetilde{\mathbf{Z}}_B = \widetilde{\mathbf{X}}_{S,K-L} = \mathbf{X}_{S,K-L}^T \mathbf{X_B}$ and $\mathbf{H}_B = \mathbf{I}$. The above assimilation problem was numerically solved with the Fletcher-Reeves-Polak-Ribiere minimization (FRPRMN) algorithm. MVAR as well as VAR described in Section 2.3 can handle nonlinear observation equations typically associated with assimilation of streamflow observations and provide full-rank solutions, as opposed to ensemble Kalman filter [17] which is optimal for linear observation equations only and provides reduced-rank solutions only. As a fix-lag smoother, both MVAR and VAR can readily account for the basin response time of fast runoff which is important to effectively capture the time-lagged effect of soil moisture to streamflow. Figure 1 presents a schematic of MVAR.

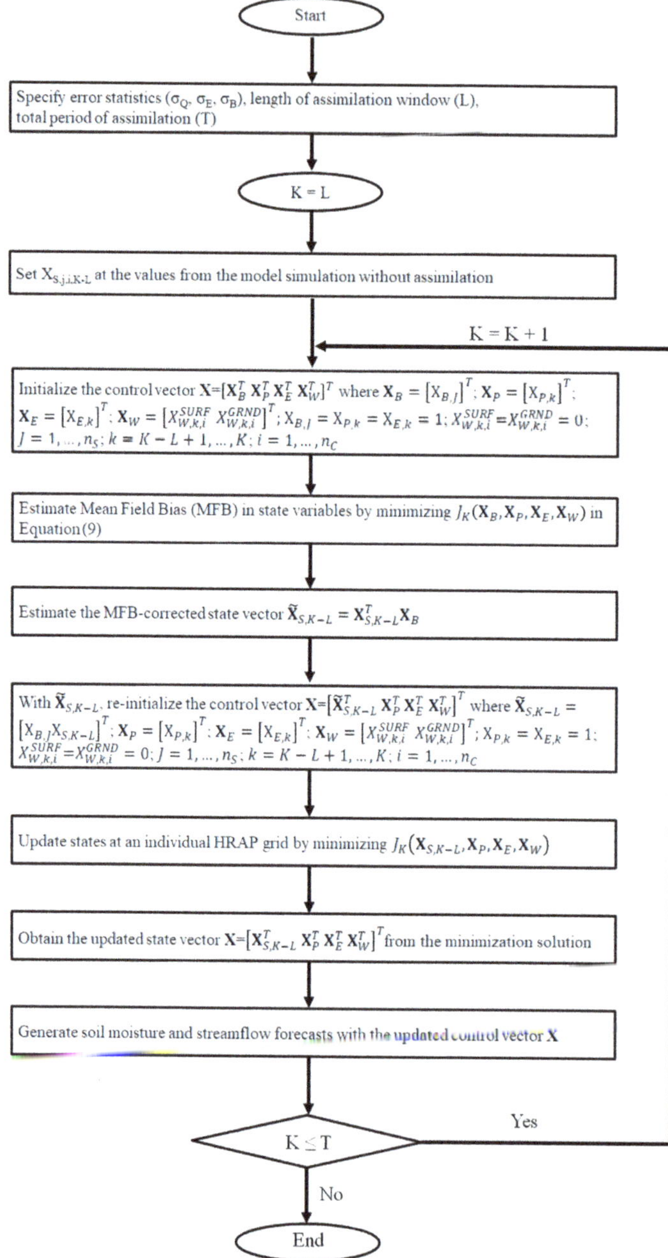

Figure 1. A schematic of the Mean Field Bias (MFB)-aware VAR (MVAR) which corrects MFB in model states via streamflow assimilation.

2.3. Conventional Variational Data Assimilation, VAR

Compared to MVAR, the conventional VAR lacks the MFB correction step and solves for model states at individual HRAP cells by minimizing Equation (12).

Minimize

$$
\begin{aligned}
J_K(\mathbf{X}_{S,K-L}, \mathbf{X}_P, \mathbf{X}_E, \mathbf{X}_W) =\ & \tfrac{1}{2}[\mathbf{Z}_S - \mathbf{H}_S \mathbf{X}_{S,K-L}]^T \mathbf{R}_B^{-1}[\mathbf{Z}_S - \mathbf{H}_S \mathbf{X}_{S,K-L}] \\
& + \tfrac{1}{2}[\mathbf{Z}_P - \mathbf{H}_P \mathbf{X}_P]^T \mathbf{R}_P^{-1}[\mathbf{Z}_P - \mathbf{H}_P \mathbf{X}_P] \\
& + \tfrac{1}{2}[\mathbf{Z}_E - \mathbf{H}_E \mathbf{X}_E]^T \mathbf{R}_E^{-1}[\mathbf{Z}_E - \mathbf{H}_E \mathbf{X}_E] \\
& + \tfrac{1}{2}\big[\mathbf{Z}_Q - \mathbf{H}_Q(\mathbf{X}_{S,K-L}, \mathbf{X}_P, \mathbf{X}_E, \mathbf{X}_W)\big]^T \mathbf{R}_Q^{-1}\big[\mathbf{Z}_Q - \mathbf{H}_Q(\mathbf{X}_{S,K-L}, \mathbf{X}_P, \mathbf{X}_E, \mathbf{X}_W)\big] \\
& + \tfrac{1}{2}\mathbf{X}_W^T \mathbf{R}_W^{-1} \mathbf{X}_W
\end{aligned}
\tag{12}
$$

subject to

$$
\begin{cases}
\mathbf{X}_{S,k} = \mathbf{M}\big(\mathbf{X}_{S,k-1}, \mathbf{X}_P, \mathbf{X}_E\big), & k = K-L+1, \ldots, K \\
X_{S,j,i}^{min} \leq X_{S,j,i,k} \leq X_{S,j,i}^{max}, & k = K-L, \ldots, K;\, j = 1, \ldots, n_S;\, i = 1, \ldots, n_C
\end{cases}
\tag{13}
$$

In Equation (12), observation equations for $\mathbf{X}_P$, $\mathbf{X}_E$, $\mathbf{X}_W$ and streamflow are identical to Equations (4) to (7). The observation equation for model states $\mathbf{X}_S$ can be written as Equation (14) in which $\mathbf{H}_S = \mathbf{I}$ and $\mathbf{Z}_S$ denotes observations of model states at $k = K - L$, that is, the beginning of the assimilation window. Due to a lack of model state observations, $\mathbf{Z}_S = \mathbf{X}_{S,K-L}$ is used.

$$
\mathbf{Z}_S = \mathbf{H}_S \mathbf{X}_{S,K-L} + \mathbf{V}_S
\tag{14}
$$

With an assumption of observational errors being independent of one another and time-invariant, Equation (12) can be rewritten as Equation (15).

Minimize

$$
\begin{aligned}
J_K(\mathbf{X}_{S,K-L}, \mathbf{X}_P, \mathbf{X}_E, \mathbf{X}_W) =\ & \tfrac{1}{2} \sum_{j=1}^{n_S} \sum_{i=1}^{n_C} \big[Z_{B,j,i,K-L} - X_{S,j,i,K-L}\big]^2 \sigma_{B,j}^{-2} \\
& + \tfrac{1}{2} \sum_{k=K-L+1}^{K} \sum_{l=1}^{n_G} \big[Z_{Q,l,k} - H_{Q,l,k}(\mathbf{X}_{S,K-L}, \mathbf{X}_P, \mathbf{X}_E, \mathbf{X}_W)\big]^2 \sigma_Q^{-2} \\
& + \tfrac{1}{2} \sum_{k=K-L+1}^{K} \sum_{i=1}^{n_C} Z_{P,i,k}^2 \big[1 - X_{P,k}\big]^2 \sigma_P^{-2} \\
& + \tfrac{1}{2} \sum_{k=K-L+1}^{K} \sum_{i=1}^{n_C} Z_{PE,i,k}^2 \big[1 - X_{E,k}\big]^2 \sigma_E^{-2}
\end{aligned}
\tag{15}
$$

subject to

$$
\begin{cases}
\mathbf{X}_{S,k} = \mathbf{M}\big(\mathbf{X}_{S,k-1}, \mathbf{X}_P, \mathbf{X}_E\big), & k = K-L+1, \ldots, K \\
X_{S,j,i}^{min} \leq X_{S,j,i,k} \leq X_{S,j,i}^{max}, & k = K-L, \ldots, K;\, j = 1, \ldots, n_S;\, i = 1, \ldots, n_C
\end{cases}
\tag{16}
$$

Same as MVAR, Equations (15) and (16) were solved by the FRPRMN algorithm. Table 2 compares the objective functions used in MVAR and VAR.

Table 2. Comparison of objective functions used in MVAR and VAR.

MFB-aware variational assimilation (MVAR)	
Step 1. Adjustment of mean field bias in model states	
Objective function	$J_K(\mathbf{X}_B, \mathbf{X}_P, \mathbf{X}_E, \mathbf{X}_W) = \frac{1}{2}[\mathbf{Z}_B - \mathbf{H}_B\mathbf{X}_B]^T \mathbf{R}_B^{-1}[\mathbf{Z}_B - \mathbf{H}_B\mathbf{X}_B]$ $+ \frac{1}{2}[\mathbf{Z}_P - \mathbf{H}_P\mathbf{X}_P]^T \mathbf{R}_P^{-1}[\mathbf{Z}_P - \mathbf{H}_P\mathbf{X}_P]$ $+ \frac{1}{2}[\mathbf{Z}_E - \mathbf{H}_E\mathbf{X}_E]^T \mathbf{R}_E^{-1}[\mathbf{Z}_E - \mathbf{H}_E\mathbf{X}_E]$ $+ \frac{1}{2}\left[\mathbf{Z}_Q - \mathbf{H}_Q(\mathbf{X}_{S,K-L}, \mathbf{X}_P, \mathbf{X}_E, \mathbf{X}_W)\right]^T \mathbf{R}_Q^{-1}\left[\mathbf{Z}_Q - \mathbf{H}_Q(\mathbf{X}_{S,K-L}, \mathbf{X}_P, \mathbf{X}_E, \mathbf{X}_W)\right]$ $+ \frac{1}{2}\mathbf{X}_W^T \mathbf{R}_W^{-1}\mathbf{X}_W$ subject to $\begin{cases} \mathbf{X}_{S,k} = \mathbf{M}(\mathbf{X}_{S,k-1}, \mathbf{X}_P, \mathbf{X}_E), \quad k = K-L+1, \dots, K \\ X_{S,j,i}^{min} \le X_{S,j,i,k} \le X_{S,j,i}^{max}, \quad k = K-L, \dots, K; j = 1, \dots, n_S; i = 1, \dots, n_C \end{cases}$
Control vector	$\mathbf{X}_B, \mathbf{X}_P, \mathbf{X}_E, \mathbf{X}_W$
Step 2. Adjustment of individual model states at each HRAP cells	
Objective function	$J_K(\widetilde{\mathbf{X}}_{S,K-L}, \mathbf{X}_P, \mathbf{X}_E, \mathbf{X}_W) = \frac{1}{2}\left[\widetilde{\mathbf{Z}}_B - \mathbf{H}_B\widetilde{\mathbf{X}}_{S,K-L}\right]^T \mathbf{R}_B^{-1}\left[\widetilde{\mathbf{Z}}_B - \mathbf{H}_B\widetilde{\mathbf{X}}_{S,K-L}\right]$ $+ \frac{1}{2}[\mathbf{Z}_P - \mathbf{H}_P\mathbf{X}_P]^T \mathbf{R}_P^{-1}[\mathbf{Z}_P - \mathbf{H}_P\mathbf{X}_P]$ $+ \frac{1}{2}[\mathbf{Z}_E - \mathbf{H}_E\mathbf{X}_E]^T \mathbf{R}_E^{-1}[\mathbf{Z}_E - \mathbf{H}_E\mathbf{X}_E]$ $+ \frac{1}{2}\left[\mathbf{Z}_Q - \mathbf{H}_Q(\mathbf{X}_{S,K-L}, \mathbf{X}_P, \mathbf{X}_E, \mathbf{X}_W)\right]^T \mathbf{R}_Q^{-1}\left[\mathbf{Z}_Q - \mathbf{H}_Q(\mathbf{X}_{S,K\ L}, \mathbf{X}_P, \mathbf{X}_E, \mathbf{X}_W)\right]$ $+ \frac{1}{2}\mathbf{X}_W^T \mathbf{R}_W^{-1}\mathbf{X}_W$ subject to $\begin{cases} \mathbf{X}_{S,k} = \mathbf{M}(\mathbf{X}_{S,k-1}, \mathbf{X}_P, \mathbf{X}_E), \quad k = K-L+1, \dots, K \\ X_{S,j,i}^{min} \le X_{S,j,i,k} \le X_{S,j,i}^{max}, \quad k = K-L, \dots, K; j = 1, \dots, n_S; i = 1, \dots, n_C \end{cases}$
Control vector	$\widetilde{\mathbf{X}}_{S,K-L}, \mathbf{X}_P, \mathbf{X}_E, \mathbf{X}_W$
Conventional variational assimilation (VAR)	
Objective function	$J_K(\mathbf{X}_{S,K-L}, \mathbf{X}_P, \mathbf{X}_E, \mathbf{X}_W) = \frac{1}{2}\left[\mathbf{Z}_B - \mathbf{H}_B\mathbf{X}_{S,K-L}\right]^T \mathbf{R}_B^{-1}\left[\mathbf{Z}_B - \mathbf{H}_B\mathbf{X}_{S,K-L}\right]$ $+ \frac{1}{2}[\mathbf{Z}_P - \mathbf{H}_P\mathbf{X}_P]^T \mathbf{R}_P^{-1}[\mathbf{Z}_P - \mathbf{H}_P\mathbf{X}_P]$ $+ \frac{1}{2}[\mathbf{Z}_E - \mathbf{H}_E\mathbf{X}_E]^T \mathbf{R}_E^{-1}[\mathbf{Z}_E - \mathbf{H}_E\mathbf{X}_E]$ $+ \frac{1}{2}\left[\mathbf{Z}_Q - \mathbf{H}_Q(\mathbf{X}_{S,K-L}, \mathbf{X}_P, \mathbf{X}_E, \mathbf{X}_W)\right]^T \mathbf{R}_Q^{-1}\left[\mathbf{Z}_Q - \mathbf{H}_Q(\mathbf{X}_{S,K-L}, \mathbf{X}_P, \mathbf{X}_E, \mathbf{X}_W)\right]$ $+ \frac{1}{2}\mathbf{X}_W^T \mathbf{R}_W^{-1}\mathbf{X}_W$ subject to $\begin{cases} \mathbf{X}_{S,k} = \mathbf{M}(\mathbf{X}_{S,k-1}, \mathbf{X}_P, \mathbf{X}_E), \quad k = K-L+1, \dots, K \\ X_{S,j,i}^{min} \le X_{S,j,i,k} \le X_{S,j,i}^{max}, \quad k = K-L, \dots, K; j = 1, \dots, n_S; i = 1, \dots, n_C \end{cases}$
Control vector	$\mathbf{X}_{S,K-L}, \mathbf{X}_P, \mathbf{X}_E, \mathbf{X}_W$

3. Study Area and Evaluation Metrics

Sections 3.1 and 3.2 describe the study area and the evaluation metrics used, respectively.

3.1. Study Area

The study area used is the 2258 km^2 headwater basin that drains into the Elk River near Tiff City, Missouri. The runoff coefficient of this basin is 0.22 based on its annual precipitation of 1117 mm and runoff of 246 mm. The basin soil consists of 39.8% silty clay, 30.5% silty clay, and 25.8% silt loam, and the basin elevation varies from 229 to 457 m (Smith et al., 2004). In addition to the United States Geological Survey (USGS) stream gauge at the outlet (USGS ID: 7189000), an interior USGS gauge at Lanagan (USGS ID: 7188885) with the drainage area of 619 km^2 is used to assess the combined effect of correcting MFB in model states and assimilating interior flow observations on updated states and streamflow. Hourly streamflow observations are available for the period of October 1992 to July 2002 for Tiff City and May 2000 to May 2006 for Lanagan. Assimilation experiment was carried out for the period of May 2000 through May 2006 using model states from base model simulation for the period of October 1992 through April 2000. In total, 15 events with observed streamflow exceeding 200 m^3/s were selected for the assimilation experiments described in Section 4. Figure 2 shows the study basin where Tiff City and Lanagan drain 135 and 35 HRAP cells, respectively. Histograms in Figure 2 show the number of HRAP cells with the same distance to stream gauges where the distance is computed based on the cell-to-cell connectivity information. As a study area of the DMIP project, further details on this basin are found at Smith et al. [39].

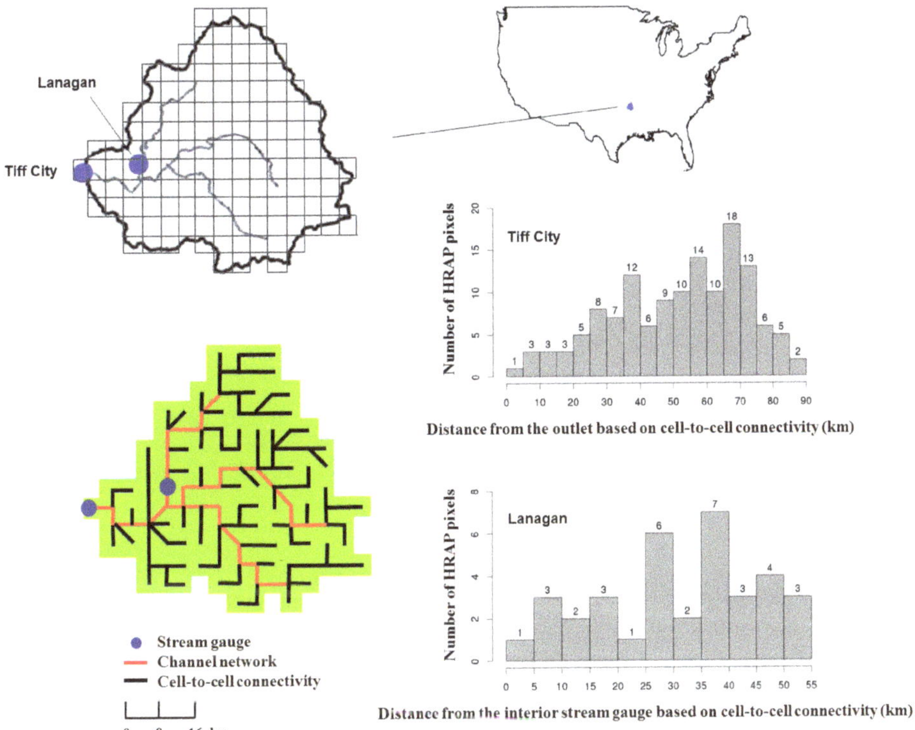

Figure 2. The TIFM7 study basin (2258 km^2); Lanagan (619 km^2) stream gauge is about 28 km from the basin outlet calculated based on the cell-to-cell connectivity. The sub-basin Lanagan consists of 35 HRAP cells (~26% drainage area of the basin).

3.2. Evaluation Metrics

To quantify changes to model states posterior to the assimilation, the Mean Absolute Difference (MAD) between a priori and updated model states at time k is computed by Equation (17).

$$MAD = \frac{1}{N} \sum_{k=1}^{N} \left| X_{S,j,i}^{+} - X_{S,j,i}^{-} \right| \tag{17}$$

In Equation (17), $X_{S,j,i}^{-}$ and $X_{S,j,i}^{+}$ represent the a priori or updated, respectively, j-th state variable at the i-th grid. In the presence of MFB in model states, the MAD values from MVAR are expected to be larger than those from VAR at HRAP cells distant to stream gauges.

Mean Squared Error (MSE) of streamflow and its decomposition into bias, variance, and co-variance terms can be calculated with Equation (18).

$$MSE = \frac{1}{N} \sum_{k=1}^{N} \left(Z_{Q,k} - H_{Q,k} \right)^2 = \left(\mu_{Z_{Q,k}} - \mu_{H_{Q,k}} \right)^2 + \left(\sigma_{Z_{Q,k}} - \sigma_{H_{Q,k}} \right)^2 + 2\sigma_{Z_{Q,k}}\sigma_{H_{Q,k}}(1-r) \tag{18}$$

where r is the linear correlation coefficient between $Z_{Q,k}$ and $H_{Q,k}$; $\mu_{Z_{Q,k}}$ and $\mu_{H_{Q,k}}$ are means of $Z_{Q,k}$ and $H_{Q,k}$, respectively; $\sigma_{Z_{Q,k}}$ and $\sigma_{H_{Q,k}}$ are standard deviations of $Z_{Q,k}$ and $H_{Q,k}$, respectively. The three components of MSE in Equation (18) help identify the source of MSE change.

To evaluate the performance of MVAR and VAR at stratified flow ranges, Type-I and Type-II Conditional Biases (CB) can be computed by Equations (19) and (20).

$$\text{Type} - \text{I CB} = Z_{Q,k} - E\left[Z_{Q,k}|H_{Q,k}\right] \tag{19}$$

$$\text{Type} - \text{II CB} = H_{Q,k} - E\left[H_{Q,k}|Z_{Q,k}\right] \tag{20}$$

Type-I CB is conditioned on streamflow simulation $H_{Q,k}$ which tells falsely detecting non-existent events or the quality of the model calibration. On the other hand, Type-II CB is conditioned on streamflow observation $Z_{Q,k}$ which quantifies a failure of detecting existing events. Both CBs can be computed by slicing the scatter plots of $Z_{Q,k}$ on an x-axis and $H_{Q,k}$ on a y-axis horizontally for Type-I CB or vertically for Type-II CB into a number of intervals. Type-I and II CBs visualize performance changes from normal to extremes. Differences between Type-II CB and Type-I CB suggest the area of further enhancement.

4. Results

In this Section, an illustrative example of MVAR and VAR results from a single assimilation cycle is presented in Section 4.1, followed by the MAD of the two assimilation techniques in terms of model states or a soil moisture profile in Section 4.2. Section 4.3 describes the effect of weakly-constrained assimilation approaches in MVAR. Section 4.4 compares the performance of MVAR and VAR on streamflow.

4.1. Illustrative Example

The sensitivity of the objective function (J_K) to background model states ($X_{S,j,i,K-L}$) at the beginning of the assimilation window, or $\frac{\partial J_K}{\partial X_{S,j,i,K-L}}$, tend to become lower with an increase of the distance to the stream gauges as shown in Figure 3 with the Normalized Mean Absolute Gradient (NMAG; Equation (21)) from all assimilation cycles.

$$NMAG = \frac{\left(\frac{1}{N}\sum_{k=1}^{N}\left|\frac{\partial J_K}{\partial X_{S,j,i,K-L}}\right|\right)}{\max\limits_{i}\left(\frac{1}{N}\sum_{k=1}^{N}\left|\frac{\partial J_K}{\partial X_{S,j,i,K-L}}\right|\right)} \tag{21}$$

In Equation (21), $X_{S,j,i,K-L}$ denotes the j-th model state at the i-th HRAP grid at $k = K - L$.

In Figure 3, TWCs show higher sensitivity than FWCs. Despite the differences in the magnitude of gradients, all five soil moisture contents in UZ and LZ show a similar spatial pattern of NMAG, that is, NMAG decrease with an increase of the distance to stream gauges. Compared to UZ states, LZ states show slightly larger NMAG around the basin outlet, indicating a weak dynamic causality between outlet flow and deep soil moisture in distant cells. Compared to the case of outlet flow assimilation, additionally assimilating interior flow increases NMAG of UZ states at the scale of both a whole basin and the sub-basin Lanagan. However, the sensitivity of LZ states has reduced in terms of a basin mean, while the opposite is true in the case of Lanagan. ADIMC is overall little sensitive to the objective function at all HRAP cells.

Normalized Mean Absolute Gradient (NMAG)

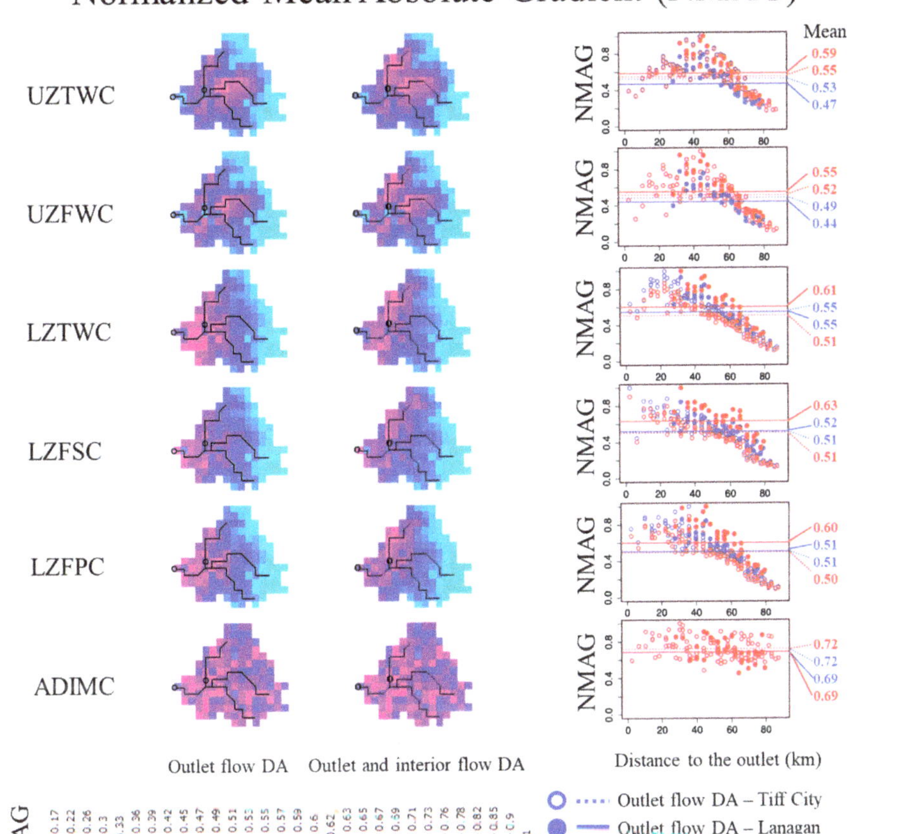

Figure 3. Time-averaged, normalized absolute gradient values of the objective function (Equation (15)) with respect to individual model states, or NMAG in Equation (21), evaluated at each HRAP cell. The largest gradient value for UZTWC, UZFWC, LZTWC, LZSFC, LZPFC, and ADIMC are 85 (86), 11 (11), 261 (301), 9 (10), 52 (58) and 0.00005 (0.00005) (mm^{-1}) in the case of assimilating outlet flow (outlet and interior flow).

Figure 4 compares MVAR with VAR for a single assimilation cycle on June 21, 15Z, 2000. At the top panel of Figure 4, MVAR adjusts states spatially uniformly to address the flow volume error at the early rising limb of Lanagan as well as Tiff City as shown in the middle panel. On the contrary, the conventional distributed state updating suffers from improving streamflow (middle panel) by adjusting individual cell states (top panel), and the change in states is overall negligible at distant HRAP cells. The changed amount in SAC states from all assimilation cycles will be described in the following section. The bottom panel of Figure 4 presents the states at the time of forecast, which show how differences in TWCs at the beginning of the assimilation window ($k = K - L$) yields differences in FWCs at the end of the assimilation window ($k = K$) as initial conditions for prediction.

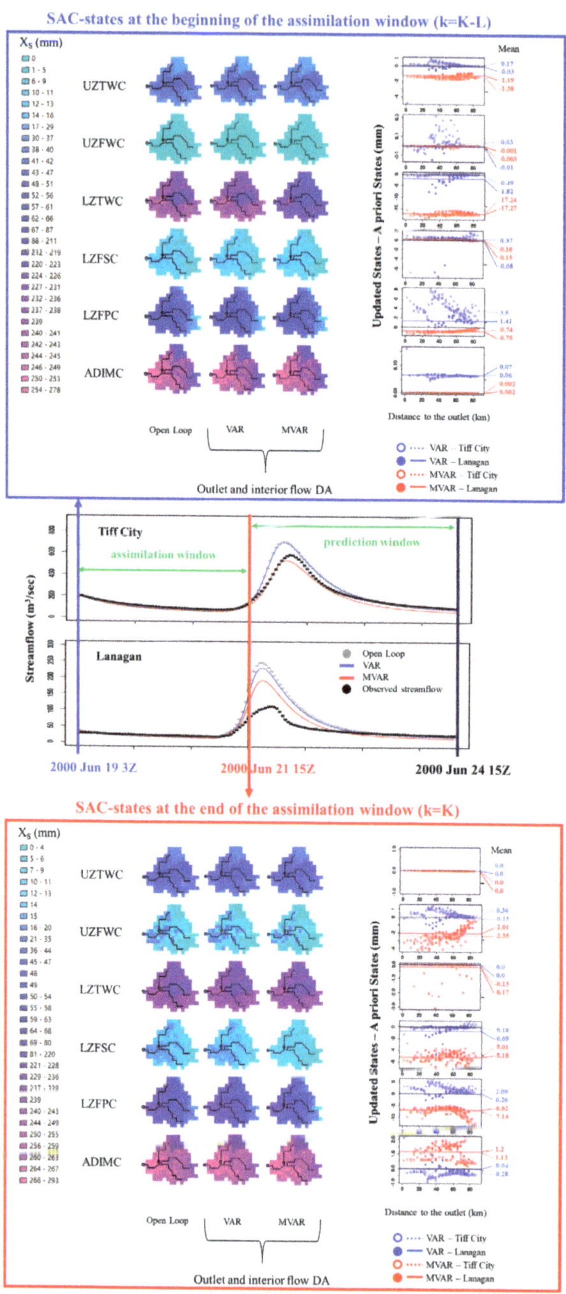

Figure 4. An example of comparing MVAR and conventional VAR results for a single assimilation cycle issued on June 21, 15Z, 2000. The sizes of assimilation and prediction windows used were 60 and 72 h. The model is applied as a strong-constraint to the assimilation problem and data assimilated includes outlet as well as interior flow observations. MFB estimates, or $X_{B,j}$ in Equation (1), are 0.9662, 1.0002, 0.9269, 1.0162, 0.9865, and 1.0 for UZTWC, UZFWC, LZTWC, LZSFC, LZPFC, and ADIMC, respectively.

4.2. Model States

Through the MFB correction, MVAR is expected to update states at distant cells more largely than VAR for all model states except ADIMC with its small gradient values everywhere (Figure 3). This is shown in Figure 5 as an increasing pattern of MAD(M)—MAD(V) with an increase of the distance to the basin outlet where MAD (Equation (17)) quantifies the mean absolute difference of a priori and updated states; MAD(M) and MAD(V) represent MAD from MVAR or VAR, respectively; LZSFC at $k = K - L$ shows less of an increasing pattern than other soil moisture states possibly due to the objective function is less sensitive to LZSFC than the other states, as shown in Figure 3. In contrast, LZTWC with the largest sensitivity shows vividly an increasing pattern at both $k = K - L$ and K, compared to the rest states. In Figure 5, MAD differences tend to decrease with an increase of NMAG particularly in the case of LZTWC, which reflects the overall decreasing pattern of NMAG with an increase of the distance (Figure 3). When assimilating interior flows in addition to outlet flows, NMAG mean values of UZ and LZ tension and free water contents for Lanagan are always larger than those for Tiff City (Figure 3). This is shown in Figure 5 as MAD differences are larger for HRAP cells within Lanagan (solid red dots) than the rest (open red circles) in the case of assimilating interior flow in addition to outlet flow. This suggests a potential need of addressing biases at a sub-basin scale, that is, adjusting Local Bias (LB) instead of MFB—left for future work. In Figure 6, soil moisture at 5, 25, 60, 75, and 100 cm depths translated from SAC states shows consistently an increasing pattern of the MAD difference with an increase of the distance to the stream gauge at both $k = K - L$ and K—this supports the effectiveness of MVAR in updating soil moisture states at distant cells despite their little sensitivities to streamflow. Compared to a single scale DA used in Lee et al. [8], MVAR is able to noticeably update model states at distant cells (Figures 5 and 6) while allowing updating model states at individual cells separately.

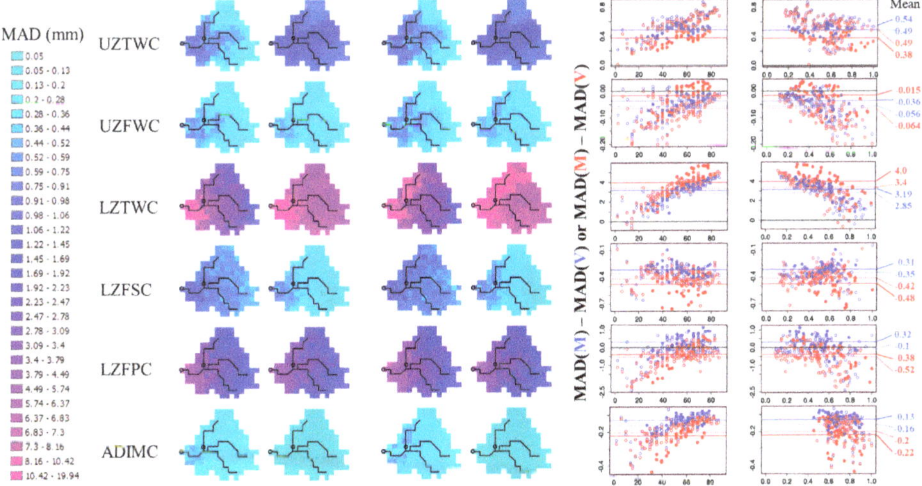

Figure 5. *Cont.*

MAD of a priori and updated SAC-states at the end of the assimilation window (k=K)

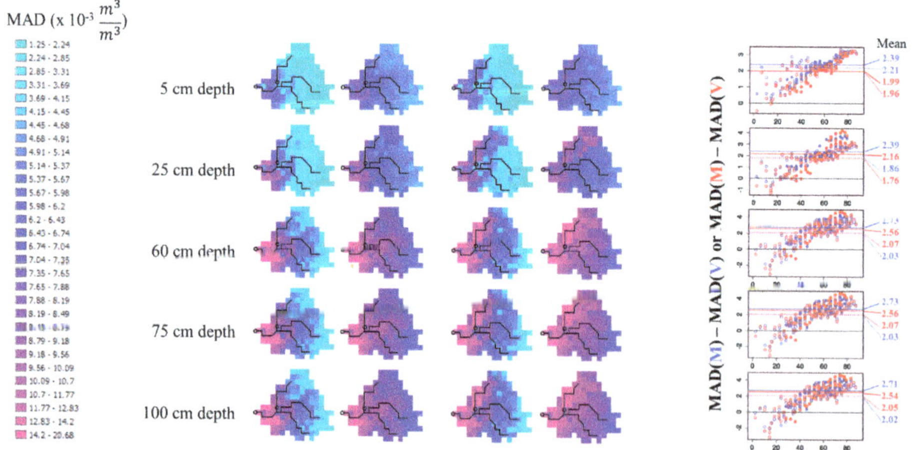

Figure 5. Mean absolute difference (MAD) of a priori and updated SAC-states at the begging ($k = K − L$; top panel) or end ($k = K$; bottom panel) of the assimilation window where the model is applied as a strong-constraint to the assimilation problem.

MAD of a priori and updated soil moisture at the beginning of the assimilation window (k=K-L)

Figure 6. *Cont.*

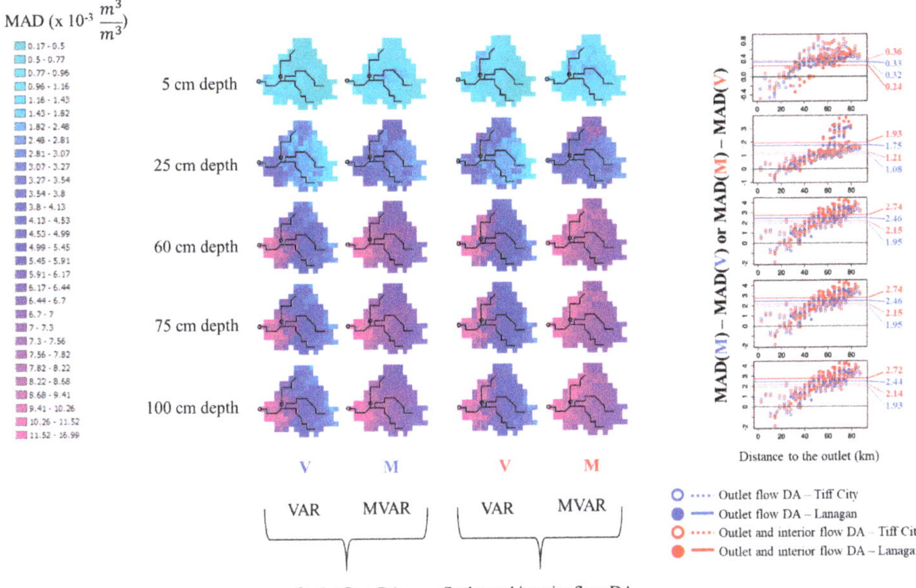

Figure 6. Same as Figure 5 but for depth-specific soil moisture translated from SAC-states via the soil moisture translation routine available in the SAC-HT.

4.3. Model Structural Error

Differences between weakly- and strongly-constrained model results may reveal the reduced amount of model structural inadequacies owing to the MFB correction in MVAR when compared to corresponding VAR results. Figure 7 indicates that, except LZTWC, accounting for MFB in model states reduces the amount of additional adjustment for SAC-states by $\mathbf{X}_W = \left[X_W^{SURF} \ X_W^{GRND} \right]^T$. The distributed SAC-SMA model may be deficient in modeling deepwater evaporation from LZ (Koren et al., 2014), which in turn causes a larger estimation of X_W^{GRND} in the case of MVAR than VAR (Figure 8), particularly at HRAP cells neighboring stream gauges. Equation (22) is used to compute $\overline{X_W^{SURF}}$ or $\overline{X_W^{GRND}}$ in Figure 8, which is the value averaged over the entire assimilation window (L) from all cycles ($N - L + 1$).

$$\overline{X_W} = \frac{1}{L(N - L + 1)} \sum_{K=L}^{N} \sum_{k=K-L+1}^{K} X_{W,k,i} \tag{22}$$

**Difference in WC and SC assimilation results
in terms of MAD of a priori and updated SAC-states at *k=K-L***

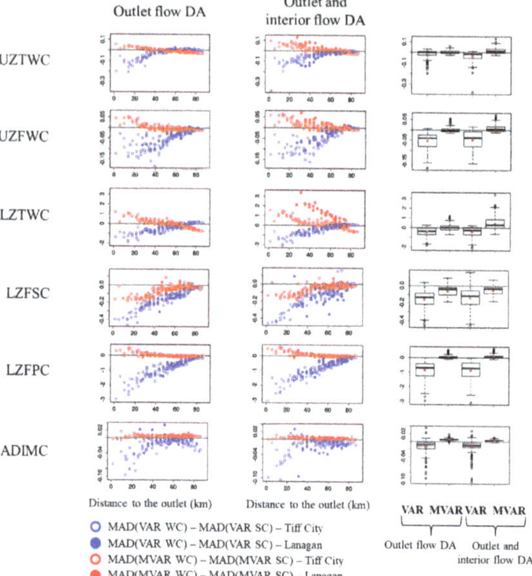

**Difference in WC and SC assimilation results
in terms of MAD of a priori and updated SAC-states at *k=K***

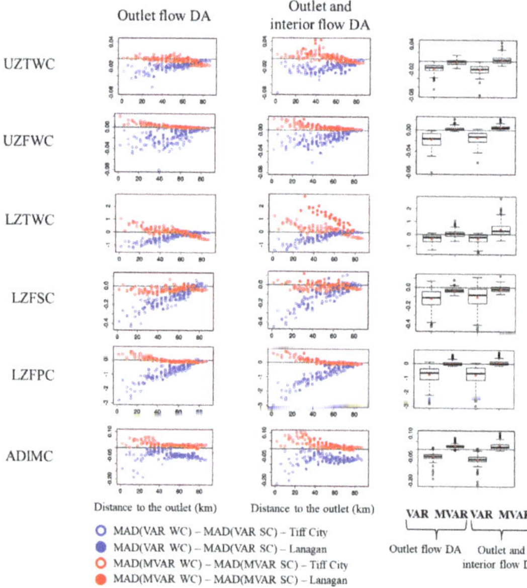

Figure 7. Differences in weakly- (WC) and strongly-constrained (SC) assimilation results in terms of MAD of background and updated model states at the beginning ($k = K − L$; top panel) or end ($k = K$; bottom panel) of the assimilation window. Box plots were created using the entire basin results, that is, Tiff City.

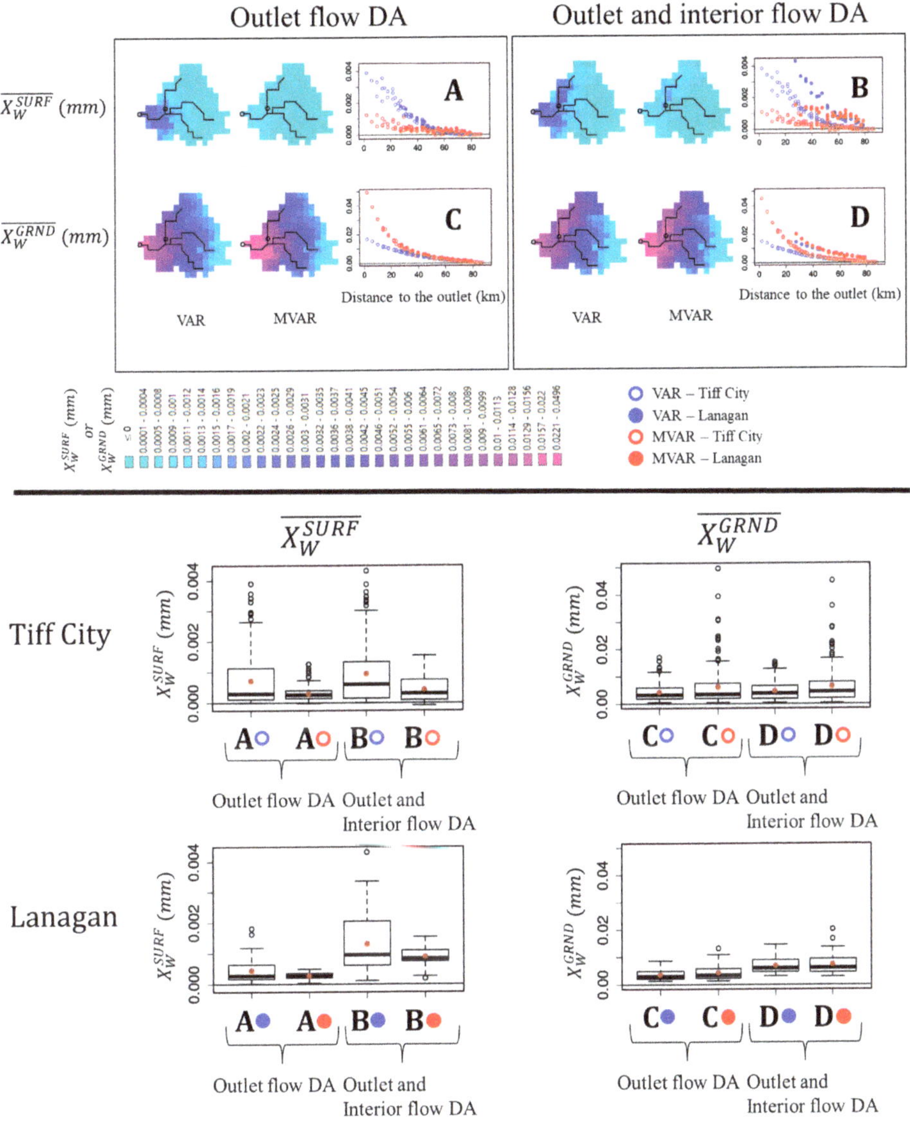

Figure 8. $\overline{X_W^{SURF}}$ and $\overline{X_W^{GRND}}$ (Equation (22)) from MVAR and VAR.

The bottom panel of Figure 8 shows box plots of $\overline{X_W^{SURF}}$ and $\overline{X_W^{GRND}}$ at the scale of the entire basin Tiff City and the sub-basin Lanagan. At both Tiff City and Lanagan, MVAR produced notably smaller mean and value range of $\overline{X_W^{SURF}}$ ($A \circ B \circ A \bullet B \bullet$ in Figure 8) than VAR ($A \circ B \circ A \bullet B \bullet$) at both flow assimilation cases. In contrast, differences in $\overline{X_W^{GRND}}$ are marginal except outliers at Tiff City ($C \circ C \circ D \circ D \circ$ at the top right box plot in Figure 8). Figures 7 and 8 suggest that, although MVAR generally reduces the effect of $\mathbf{X}_W$ on model states by possibly reducing the degree of compensating for unresolved MFB by $\mathbf{X}_W$, improving model fidelity to physical realism may precede an assimilation procedure in order to pose the assimilation problem well-posed [34]. When assimilating interior flows in addition to outlet flow, $\overline{X_W^{SURF}}$ and $\overline{X_W^{GRND}}$ values within Lanagan ($\bullet$ and $\bullet$ at the top panel of Figure 8) are conspicuously different from the rest (∘ and ∘). This may indicate the changing amount of model structural errors as well as field biases resolvable via additionally assimilating interior flows which may benefit from adjusting LB at the scale of Lanagan, as discussed in Section 4.2. Table 3 shows the time-averaged spatial correlation of background and updated model states from both MVAR and VAR. In Table 3, MVAR generally retains higher spatial correlations than VAR at most states at all assimilation cases at both $k = K - L$ and K. When the model is applied as a weak-constraint, MVAR-generated LZTWC shows consistently lower spatial correlations than the VAR-generated. As discussed above, this is also an indication of a need to improve model physics prior to the assimilation in order to preserve the spatial correlation structure of background model states as driven by base model simulation [34].

Table 3. Time-averaged spatial correlation of background and updated model states. Values in the parenthesis represent correlation from VAR, or r_V, otherwise from MVAR, or r_M. Bold font highlights a larger value between r_V and r_M. Underscored is the pair showing the minimum or maximum value of $r_M - r_V$ at either $k = K - L$ or $k = K$. WC and SC represent the model applied as a weak- or strong-constraint to the assimilation problem, respectively.

	UZTWC	UZFWC	LZTWC	LZSFC	LZPFC	ADIMC
	Begging of the assimilation window (k = K − L)					
	Outlet flow assimilation					
WC	**0.995**(0.991)	**0.96**(0.876)	0.88(**0.883**)	**0.964**(0.871)	**0.981**(0.96)	0.999(0.999)
SC	**0.996**(0.99)	**0.969**(0.853)	**0.902**(0.863)	**0.966**(0.854)	**0.993**(0.95)	0.999(0.999)
	Outlet and interior flow assimilation					
WC	**0.992**(0.99)	**0.952**(0.859)	0.817(**0.838**)	**0.956**(0.828)	**0.977**(0.944)	**0.999**(0.998)
SC	**0.994**(0.989)	**0.963**(0.837)	**0.854**(0.817)	**0.961**(0.823)	**0.99**(0.935)	**0.999**(0.998)
	End of the assimilation window (k = K)					
	Outlet flow assimilation					
WC	0.999(0.999)	**0.845**(0.765)	0.913(**0.97**)	**0.872**(0.85)	**0.947**(0.932)	**0.997**(0.996)
SC	0.999(0.999)	**0.842**(0.752)	0.931(**0.951**)	**0.883**(0.841)	**0.959**(0.926)	**0.997**(0.995)
	Outlet and interior flow assimilation					
WC	0.998(**0.999**)	**0.821**(0.754)	0.88(**0.966**)	**0.814**(0.768)	**0.928**(0.909)	**0.996**(0.994)
SC	0.999(0.999)	**0.834**(0.733)	0.912(**0.945**)	**0.831**(0.768)	**0.943**(0.901)	**0.997**(0.994)

4.4. Streamflow

MVAR improved outlet as well as interior flow over the conventional VAR at all assimilation cases in terms of MSE (Figure 9). Compared to VAR, MVAR reduced further MSE of streamflow by 2–8% at Tiff City, and by 1–10% at Lanagan, depending on the data assimilated and the model applied as a weak- or strong-constraint to the assimilation problem (Figure 9). Assimilating interior flow in addition to outlet flow is necessary to achieve considerable improvement in interior flows by VAR [8], which is further improved by MVAR. Compared to strongly-constrained assimilation results, weakly-constrained assimilation reduced MSE of streamflow by 2–4% at Tiff City and by 0.4–3% at Lanagan. MSE decomposition into bias, variance, and co-variance indicates that MVAR outperforms VAR by better modeling co-variabilities of streamflow observation and simulation at both Tiff City and Lanagan at the cost of increasing bias and variance terms at some assimilation cases. In the case of assimilating outlet flow only, all three MSE components for outlet flow are consistently reduced by MVAR more than VAR, implying the positive effect of addressing MFB in model states on reducing bias and modeling variance as well as co-variance in simulated and observed flow. Magnitude-dependent performance is examined based on Type-I and –II CBs. In Figure 10, Type-I CB is generally smaller than Type-II CB particularly at extremes which signifies the importance of addressing Type-II CB in estimation and prediction of extremes [23]. In the case of assimilating interior flows in addition to outlet flow, median-range flows of less than 150 m^3/s at Lanagan are mostly improved, whereas heavy-to-extreme flows at Lanagan are degraded (Figure 10). Visual examination of hydrographs at Lanagan indicated the notable amount of magnitude-dependent flow bias at some events which warrants future efforts with CB-penalized assimilation techniques such as CB-penalized Kalman Filter (CBPKF, [23]) or its ensemble extension, CB-penalized Ensemble Kalman Filter (CBEnKF, [9]).

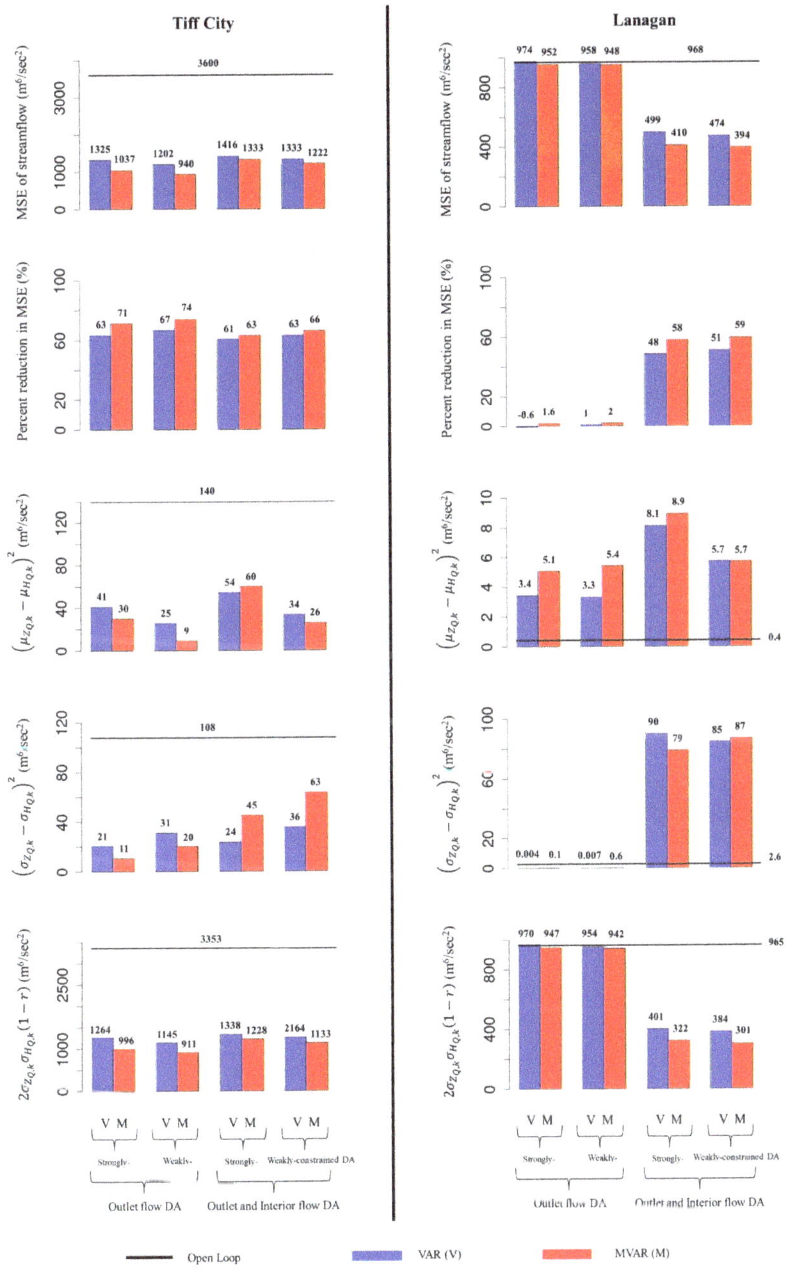

Figure 9. Mean Squared Error (MSE) of streamflow analysis generated from VAR (V) or MVAR (M) and its decomposition (Equation (18)) where $\mu_{Z_{Q,k}}$ and $\sigma_{Z_{Q,k}}$ are 125 (m^3/s) and 157 (m^3/s) for Tiff City, and 30 (m^3/s) and 43 (m^3/s) for Lanagan.

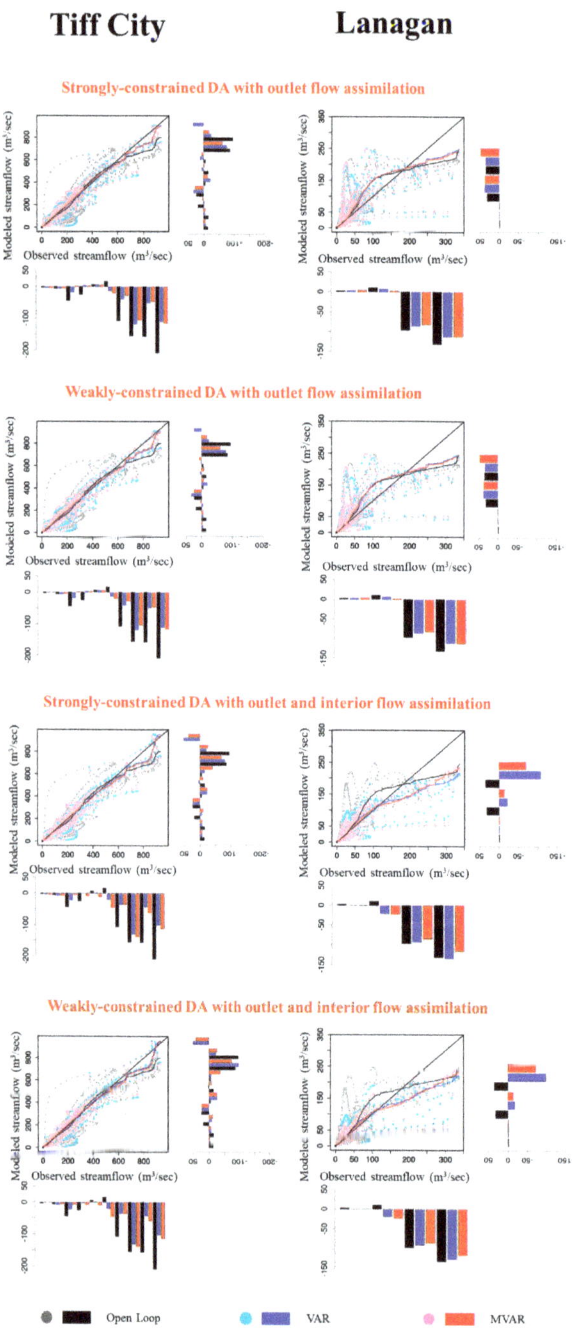

Figure 10. Scatter plots of observed and simulated flow from the open loop, VAR, and MVAR at Tiff City and Lanagan. Type-I (Equation (19)) and Type–II (Equation (20)) conditional biases in streamflow are shown as bar plots on the right-hand side or at the bottom, respectively, of the scatter plot.

5. Conclusions

The MFB-aware variational assimilation (MVAR) for the distributed SAC-SMA model was developed and comparatively evaluated with the conventional VAR based on its application to the headwater basin at Tiff City, Missouri. MVAR corrects MFB in model states and then update states at an individual HRAP grid. Compared to VAR [7,8,11,34], MVAR adjusts model states at remote cells by larger margins albeit their little sensitivities to streamflow. MVAR generally outperformed VAR in improving streamflow in both cases of outlet flow assimilation and interior as well as outlet flow assimilation. When the interior flow is assimilated in addition to outlet flow assimilation, improvement in interior flow is conspicuous. When the model is applied as a weak-constraint to the assimilation problem, MVAR generally less adjusted model states than VAR, implying the model error estimate possibly compensated for the MFB in model states in the case of VAR. The future work includes developing VAR with a capability to address spatially non-uniform biases in model states as well as magnitude-dependent biases in streamflow.

Author Contributions: Conceptualization, H.L., H.S. and D.-J.S.; methodology, H.L., D.-J.S.; software, H.L. and D.-J.S.; validation, H.L., H.S. and D.-J.S.; formal analysis, H.L.; investigation, H.L., H.S. and D.-J.S.; writing—original draft preparation, H.L.; writing—review and editing, H.L., H.S. and D.-J.S.; supervision, D.-J.S.; project administration, H.L. and D.-J.S.; funding acquisition, D.-J.S. All authors have read and agreed to the published version of the manuscript.

Funding: The second and third authors were supported in part by the National Oceanic and Atmospheric Administration's Joint Technology Transfer Initiative Program (Grants NA16OAR4590232, NA17OAR4590174, and NA17OAR4590184). These supports are gratefully acknowledged.

Conflicts of Interest: The authors declare no conflict of interest.

References

1. Kitanidis, P.K.; Bras, R.L. Real time forecasting of river flows. R. M. Parsons laboratory for water resources and hydrodynamics. *Tech. Rep.* **1978**, *235*, 324.
2. Sittner, W.T.; Krouse, K.M. *Improvement of Hydrologic Simulation by Utilizing Observed Discharge as an Indirect Input (Computed Hydrograph Adjustment Technique—CHAT)*; NOAA Tech. Memo. NWS HYDRO-38: Silver Spring, MD, USA, 1979; p. 125.
3. WMO. *Simulated Real-Time Intercomparison of Hydrological Models. Operational Hydrology Rep. 38*; World Meteorological Organization: Geneva, Switzerland, 1992; p. 241.
4. Georgakakos, K.; Smith, G.F. On improved operational hydrologic forecasting: Results from a WMO real-time forecasting experiment. *J. Hydrol.* **1990**, *114*, 17–45. [CrossRef]
5. Refsgaard, J.C. Validation and intercomparison of different updating procedures for real-time forecasting, Nord. *Hydrol. Res.* **1998**, *28*, 65–84. [CrossRef]
6. Moradkhani, H.; Hsu, K.-L.; Gupta, H.; Sorooshian, S. Uncertainty assessment of hydrologic model states and parameters: Sequential data assimilation using the particle filter. *Water Resour. Res.* **2005**, *41*, W05012. [CrossRef]
7. Lee, H.; Seo, D.-J.; Koren, V. Assimilation of streamflow and in-situ soil moisture data into operational distributed hydrologic models: Effects of uncertainties in the data and initial model soil moisture states. *Adv. Water Resour.* **2011**, *34*, 1597–1615. [CrossRef]
8. Lee, H.; Seo, D.-J.; Liu, Y.; Koren, V.; McKee, P.; Corby, R. Variational assimilation of streamflow into operational distributed hydrologic models: Effect of spatiotemporal scale of adjustment. *Hydrol. Earth Syst. Sci.* **2012**, *16*, 2233–2251. [CrossRef]
9. Lee, H.; Shen, H.; Noh, S.J.; Kim, S.; Seo, D.-J.; Zhang, Y. Improving flood forecasting using conditional bias-penalized ensemble Kalman filter. *J. Hydrol.* **2019**, *575*, 596–611. [CrossRef]
10. Rakovec, O.; Weerts, A.H.; Hazenberg, P.; Torfs, P.J.J.F.; Uijlenhoet, R. State updating of a distributed hydrological model with Ensemble Kalman Filtering: Effects of updating frequency and observation network density on forecast accuracy. *Hydrol. Earth Syst. Sci.* **2012**, *16*, 3435–3449. [CrossRef]

11. Lee, H.; Seo, D.-J. Assimilation of hydrologic and hydrometeorological data into distributed hydrologic model: Effect of adjusting mean field bias in radar-based precipitation estimates. *J. Hydrol.* **2014**, *74*, 196–211. [CrossRef]

12. Nasab, A.R.; Seo, D.-J.; Lee, H.; Kim, S. Comparative evaluation of maximum likelihood ensemble filter and ensemble Kalman filter for real-time assimilation of streamflow data into operational hydrologic models. *J. Hydrol.* **2014**, *519*, 2663–2675.

13. Dooge, J.C.I.; Napiôrkowsld, J.J. Applicability of diffusion analogy in flood routing. *Acta Geophys. Pol.* **1987**, *35*, 65–75.

14. Jia, X.; Zeng, F.; Gu, Y. Semi-analytical solutions to one-dimensional advection–diffusion equations with variable diffusion coefficient and variable flow velocity. *Appl. Math. Comput.* **2013**, *221*, 268–281. [CrossRef]

15. Nazari, B.; Seo, D.-J. Symbolic explicit solutions for 1-dimensional linear diffusive wave equation with lateral inflow and their applications. *Water Resour. Res.* **2020**. [CrossRef]

16. Li, Z.; Navon, I.M. Optimality of variational data assimilation and its relationship with the Kalman filter and smoother. *Quart. J. Roy. Meteor. Soc.* **2001**, *127*, 661–683. [CrossRef]

17. Evensen, G. Sequential data assimilation with a nonlinear quasi-geostrophic model using Monte Carlo methods to forecast error statistics. *J. Geophys. Res.* **1994**, *99*, 10143–10162. [CrossRef]

18. De Lannoy, G.J.M.; Reichle, R.H.; Houser, P.R.; Pauwels, V.R.N.; Verhoest, N.E.C. Correcting for forecast bias in soil moisture assimilation with the ensemble Kalman filter. *Water Resour. Res.* **2007**, *43*, W09410. [CrossRef]

19. Reichle, R.H.; Koster, R.D. Bias reduction in short records of satellite soil moisture. *Geophys. Res. Lett.* **2004**, *31*, L19501. [CrossRef]

20. Seo, D.-J.; Breidenbach, J.P. Real-time correction of spatially nonuniform bias in radar rainfall data using rain gauge measurements. *J. Hydrometeor.* **2002**, *3*, 93–111. [CrossRef]

21. Wu, L.; Seo, D.-J.; Demargne, J.; Brown, J.D.; Cong, S.; Schaake, J. Generation of ensemble precipitation forecast from single-valued quantitative precipitation forecast for hydrologic ensemble prediction. *J. Hydrol.* **2011**, *399*, 281–298. [CrossRef]

22. Seo, D.-J.; Herr, H.D.; Schaake, J.C. A statistical post-processor for accounting of hydrologic uncertainty in short-range ensemble streamflow prediction. *Hydrol. Earth Syst. Sci. Discuss.* **2006**, *3*, 1987–2035. [CrossRef]

23. Seo, D.-J.; Saifuddin, M.M.; Lee, H. Conditional bias-penalized Kalman filter for improved estimation and prediction of extremes. *Stoch. Environ. Res. Risk Assess.* **2018**, *32*, 183–201. [CrossRef]

24. Chepurin, G.A.; Carton, J.A.; Dee, D. Forecast model bias correction in ocean data assimilation. *Mon. Wea. Rev.* **2005**, *133*, 1328–1342. [CrossRef]

25. Dee, D.P. Bias and data assimilation. *Q. J. R. Meteorol. Soc.* **2005**, *131*, 3323–3343. [CrossRef]

26. Brown, J.D.; Wu, L.; He, M.; Regonda, S.; Lee, H.; Seo, D.-J. Verification of temperature, precipitation and streamflow forecasts from the NOAA/NWS Hydrologic Ensemble Forecast Service (HEFS): 1. Experimental design and forcing verification. *J. Hydrol.* **2014**, *519*, 2869–2889. [CrossRef]

27. Brown, J.D.; He, M.; Regonda, S.; Wu, L.; Lee, H.; Seo, D.-J. Verification of temperature, precipitation and streamflow forecasts from the NOAA/NWS hydrologic ensemble forecast service (HEFS): 1. Streamflow verification. *J. Hydrol.* **2014**, *519*, 2847–2868. [CrossRef]

28. Smith, J.A.; Krajewski, W.F. Estimation of the mean field bias of radar rainfall estimates. *J. Appl. Meteorol.* **1991**, *30*, 397–412. [CrossRef]

29. Seo, D.-J.; Breidenbach, J.P.; Johnson, E.R. Real-time estimation of mean field bias in radar rainfall data. *J. Hydrol.* **1999**, *223*, 131–147. [CrossRef]

30. Kim, T.-J.; Kwon, H.-H.; Lima, C. A Bayesian partial pooling approach to mean field bias correction of weather radar rainfall estimates: Application to Osungsan weather radar in South Korea. *J. Hydrol.* **2018**, *565*, 14–26. [CrossRef]

31. Koren, V.; Reed, S.; Smith, M.; Zhang, Z.; Seo, D.-J. Hydrology laboratory research modeling system (HL-RMS) of the US national weather service. *J. Hydrol.* **2004**, *291*, 297–318. [CrossRef]

32. Li, Z.; McWilliams, J.C.; Ide, K.; Farrara, J.D. A multiscale variational data assimilation scheme: Formulation and illustration. *Mon. Wea. Rev.* **2015**, *143*, 3804–3822. [CrossRef]

33. Noh, S.J.; Weerts, A.H.; Rakovec, O.; Lee, H.; Seo, D.-J. Assimilation of streamflow observations. In *Handbook of Hydrometeorological Ensemble Forecasting*; Duan, Q., Pappenberger, F., Thielen, J., Wood, A., Cloke, H.L., Schaake, J.C., Eds.; Springer: Berlin/Heidelberg, Germany, 2018; pp. 1–36. [CrossRef]

34. Lee, H.; Seo, D.-J.; Noh, S.J. A weakly-constrained data assimilation approach to address rainfall-runoff model structural inadequacy in streamflow prediction. *J. Hydrol.* **2016**, *542*, 373–391. [CrossRef]

35. Burnash, R.J.; Ferral, R.L.; McGuire, R.A. *A Generalized Streamflow Simulation System: Conceptual Modeling for Digital Computers*; US Department of Commerce National Weather Service and State of California Department of Water Resources: Sacramento, CA, USA, 1973.

36. Reed, S.M.; Maidment, D.R. Coordinate transformations for using NEXRAD data in GIS-based hydrologic modeling. *J. Hydrol. Eng.* **1999**, *4*, 174–183. [CrossRef]

37. Reed, S.M. Deriving flow directions for coarse-resolution (1–4 km) gridded hydrologic modeling. *Water Resour. Res.* **2003**, *39*, 1238. [CrossRef]

38. Natural Resources Conservation Service, United States Department of Agriculture, US General Soil Map (STATSGO2). Available online: https://data.nal.usda.gov/dataset/united-states-general-soil-map-statsgo2 (accessed on 10 December 2020).

39. Smith, M.B.; Seo, D.-J.; Koren, V.I.; Reed, S.M.; Zhang, Z.; Duan, Q.; Moreda, F.; Cong, S. The distributed model intercomparison project (DMIP): Motivation and experiment design. *J. Hydrol.* **2004**, *298*, 4–26. [CrossRef]

40. Koren, V.; Smith, M.; Cui, Z. Physically-based modifications to the sacramento soil moisture accounting model. Part A: Modeling the effects of frozen ground on the runoff generation process. *J. Hydrol.* **2014**, *519*, 3475–3491. [CrossRef]

41. Sasaki, Y. Some basic formalisms in numerical variational analysis. *Mon. Weather. Rev.* **1970**, *98*, 875–883. [CrossRef]

42. Zupanski, D. A general weak constraint applicable to operational 4DVAR data assimilation systems. *Mon. Weather. Rev.* **1997**, *125*, 2274–2292. [CrossRef]

43. Di Lorenzo, E.; Moore, A.M.; Arango, H.G.; Cornuelle, B.D.; Miller, A.J.; Powell, B.; Chua, B.S.; Bennett, A.F. Weak and strong constraint data assimilation in the inverse regional ocean modeling system (ROMS): Development and application for a baroclinic coastal upwelling system. *Ocean Model.* **2007**, *16*, 160–187. [CrossRef]

44. Moradkhani, H.; Sorooshian, S.; Gupta, H.V.; Houser, P.R. Dual state-parameter estimation of hydrological models using ensemble Kalman filter. *Adv. Water Resour.* **2005**, *28*, 135–147. [CrossRef]

45. Ravela, S.; Emanuel, K.; McLaughlin, D. Data assimilation by field alignment. *Phys. D Nonlinear Phenom.* **2007**, *230*, 127–145. [CrossRef]

46. Beven, K.J. On the concept of model structural error. *Water Sci. Technol.* **2005**, *52*, 167–175. [CrossRef] [PubMed]

47. Seo, D.-J.; Koren, V.; Cajina, N. Real-time variational assimilation of hydrologic and hydrometeorological data into operational hydrologic forecasting. *J. Hydrometeorol.* **2003**, *4*, 627–641. [CrossRef]

Publisher's Note: MDPI stays neutral with regard to jurisdictional claims in published maps and institutional affiliations.

 *forecasting*

Article

Forecasting of Future Flooding and Risk Assessment under CMIP6 Climate Projection in Neuse River, North Carolina

Indira Pokhrel [1], Ajay Kalra [1,*](image), Md Mafuzur Rahaman [2] and Ranjeet Thakali [3]

[1] School of Civil, Environmental and Infrastructure Engineering, Southern Illinois University, 1230 Lincoln Drive, Carbondale, IL 62901-6603, USA; indira.pokhrel@siu.edu

[2] AECOM, 2380 McGee St Suite 200, Kansas City, MO 64108, USA; mafuzur.rahaman@aecom.com

[3] Bayer-Risse Engineering, Inc., 78 State Highway 173 W, Suite#6, Hampton, NJ 08827, USA; rthakali@bayer-risse.com

* Correspondence: kalraa@siu.edu; Tel.: +1-(618)-453-7008

Received: 7 July 2020; Accepted: 18 August 2020; Published: 28 August 2020

Abstract: Hydrological extremes associated with climate change are becoming an increasing concern all over the world. Frequent flooding, one of the extremes, needs to be analyzed while considering climate change to mitigate flood risk. This study forecast streamflow and evaluate risk of flooding in the Neuse River, North Carolina considering future climatic scenarios, and comparing them with an existing Federal Emergency Management Agency study. The cumulative distribution function transformation method was adopted for bias correction to reduce the uncertainty present in the Coupled Model Intercomparison Project Phase 6 (CMIP6) streamflow data. To calculate 100-year and 500-year flood discharges, the Generalized Extreme Value (L-Moment) was utilized on bias-corrected multimodel ensemble data with different climate projections. Out of all projections, shared socio-economic pathways (SSP5-8.5) exhibited the maximum design streamflow, which was routed through a hydraulic model, the Hydrological Engineering Center's River Analysis System (HEC-RAS), to generate flood inundation and risk maps. The result indicates an increase in flood inundation extent compared to the existing study, depicting a higher flood hazard and risk in the future. This study highlights the importance of forecasting future flood risk and utilizing the projected climate data to obtain essential information to determine effective strategic plans for future floodplain management.

Keywords: streamflow; CMIP6; bias correction; HEC-RAS; flood inundation maps; risk assessment

---

1. Introduction

According to the Intergovernmental Panel on Climate Change (IPCC) report, climate change has triggered extreme hydrologic events that are having adverse effect on the hydrological cycle and human livelihoods all over the world [1]. In addition, climate change plays a vital role in the ecosystem shift, leading to an increase in global warming [2,3]. Moreover, greenhouse gas emissions (GHG), induced by anthropogenic factors, are escalating the rate of global warming. If the rate of GHG emissions continues to increase at the same rate as today, others predict global warming to increase from 1.5 to 2 °C by 2052 [1]. The rising trend in global warming maximizes the evaporation rate of surface water and soil moisture that will affect the amount of precipitation all around the globe [4]. This type of alteration in the hydrologic cycle will affect the runoff and availability of both the surface and subsurface water, which eventually impacts river streamflow [5]. Furthermore, the regional precipitation trend will be affected by the patterns of ocean currents and wind, which will ultimately result in a change in streamflow. Subsequently, high flows and low flows i.e., floods and droughts, are likely to occur more frequently across the globe with increased intensity [6–8]. These increases in both high and low flow extremes have already occurred in some parts of the world and are making societal infrastructure more sensitive to climate change [3]. Moreover, extreme events are more frequent these days and are anticipated to continue at the same frequency or even faster in the future.

Flooding is one of the most frequently occurring natural hazards in the world, along with droughts and heatwaves [7,8]. Due to the changing intensity and frequency of extreme rainfall, different parts of the world are facing deadly flood events causing a significant loss of life and property [9]. Some studies [7,10] analyze the potential impact of climate change and flood risk on a global scale, where they discuss the change in flood frequency and its impact on population centers in different parts of the world. In the United States, various studies [11–13] evaluated the change in regional streamflow due to the changes in climate. Arnell et al. [14] showed that the alterations in annual precipitation distribution in North Carolina (NC) with the warming climate might increase extreme flooding events and water quality problems in the future. Moreover, global warming is a threatening issue for a coastal state like NC. As per the IPCC [1], rising sea levels due to global climate change pose a high risk for coastal communities and low-lying areas, as there will be a higher possibility of consequences such as high tides, storms, and flooding. Johnson et al. [12] predicted, under different future climate change scenarios, that the Neuse River Basin (NRB) in NC will experience an increase in streamflow due to the increase in rainfall intensity. In recent years, NRB has endured deadly flood hazards due to heavy rainfalls resulted in from hurricanes Matthew in 2016 and Florence in 2018. Hall [15] estimated annual losses of USD 54 billion across the United States due to the extreme flooding events induced by the hurricane and tropical storms [15]. The National Flood Insurance Program was established in the United States in 1968 and managed by the Federal Emergency Management Agency (FEMA) to provide quick response to reduce the impact of flooding. Flood mapping is one of many programs organized by FEMA in various areas that are more vulnerable to flooding in the USA. FEMA utilizes software such as the HEC-RAS, to perform one-dimensional (1D) hydraulic modeling for the generation of flood inundation maps, which can also be used for flood risk analysis and insurance programs [16].

HEC-RAS, a hydraulic analysis program, was developed by the US Army Corps of Engineers in 1995 to work in a network that would support multi-user and multi-tasking environments. HEC-RAS helps to simulate water surface profiles for steady and unsteady flow, water quality analysis, and sediment transport computation, utilizing a graphical User Interface for interaction with the system [17,18]. Additionally, using the RAS Mapper, HEC-RAS can generate inundation maps and analyze flood patterns. The outcome of HEC-RAS simulation can be used for floodplain mapping, management, and insurance studies. Many researchers have applied the HEC-RAS 1D steady analysis for the simulation of a floodplain flows [19–23]. Some researchers extended the study to the hazard, vulnerability, and risk assessment of floodplains using geographic information systems, concluding that this is an important step for flood risk management [24–28]. Tingsanchali and Karim [25] deduced that the impact of future flooding could be identified using a risk assessment to mitigate human losses and attenuate economic as well as environmental losses. Thus, flood hazards, vulnerability, and risk assessments provide a framework for the management of flood risk. Furthermore, Noren et al. [29] suggested that a risk assessment is a vital step in effective flood risk management for a sustainable livelihood and agricultural system management. The risk zone maps can be used as source information to prepare emergency response plans, flood management and prevention programs, and infrastructure design. The current study provides risk analysis and mapping to improve early responses to future floods.

The Coupled Model Intercomparison Project (CMIP) was started in 1995 under the World Climate Research Program to evaluate the change in the climate from past to future in the multimodel context alongside the change in radioactive forcing and natural, unforced variability [30,31]. The CMIP helps researchers and policymakers to access the impact of climate change on hydro-climatological variables such as precipitation, temperature, and streamflow using different climate projections [31]. In the sixth phase of CMIP, CMIP6, climate projections were driven by scenarios based on shared socio-economic pathways (SSP). The SSPs are updated and revised versions of representative concentration pathways

(RCPs) [32], which include anthropogenic drivers such as future emissions and land use, along with socio-economic development [33,34]. Hence, in ScenarioMIP, the primary activity of CMIP6, updated scenarios share a matrix of SSP and RCPs as an SSP_{x-y}, in which x represents the specific SSP and y represents the forcing pathways from RCP. In CMIP6 climate models, based on the observation data, and historical Atmosphere–Ocean Coupled General Circulation Model (AOGCM) data obtained from different global climate models (GCMs) used future climate projections for different scenarios [32]. GCMs provide the resources to understand potential changes in regional and global climates [35–37]. Since the ensemble climate projection data improve the uncertainty and model biases, the literature recommends that there are at least two or more GCMs and multi-members of the GCMs for a multimodel ensemble to investigate the impact of future extremes [38–42]. Different GCMs were utilized by numerous studies [6,7,43] to discuss the various climate change simulations, considering the extreme daily precipitation and temperature. However, there are limited studies [8,11,44] related to the future projection of streamflow. For different recurrence intervals, the future streamflow was estimated in the current study utilizing CMIP6 streamflow projections from multiple GCMs.

GCMs of different climate projections encompass different systematic error and uncertainty factors called biases, which need to be corrected to rectify the over- or underestimated results [43,45,46]. Different studies have suggested different bias correction techniques to maintain consistency in future projection and reduce the error in GCMs outputs that would help to analyze the impact of climate change [47–53]. A cumulative distribution function- transformation (CDF-t) bias correction technique was developed by Michelangeli as the extension of quantile-matching that directly deals with and provide CDFs [50,51]. The theory of CDF-t assumes that the translation of CDF of GCMs variables is only possible where the transformation function exists [52]. It was utilized by many researchers to effectively reduce the bias for the given large-scale future hydroclimatic data [50–53]. Furthermore, Yuan and Wood [54] concluded that bias correction of streamflow was more efficient and reliable for ensemble forecasts. In the current study, bias-corrected multimodel GCMs is employed for the evaluation of peak discharge events.

Among the different probability distributions in the past studies [44,55–58], Generalized Extreme Value (GEV) was considered as a more efficient and better fit for the streamflow distribution. GEV is also used as a statistical best fit distribution to predict future extremes, especially hydroclimatic extremes by several researchers [55,59]. GEV is a parametric distribution, which utilizes shape, location, and scale parameters to find the cumulative probability for the given event [60]. Previous studies by Hosking and Wallis [60,61] suggested L-moment as one of the methods that was widely used for the estimation of the three parameters used in this distribution. GEV is commonly used in the streamflow distribution of the humid subtropical regions [59,62–64]. Additionally, in the current study, GEV was employed for the estimation of design flows at different recurrence intervals. The study methodology further uses the peak annual streamflow obtained from GEV to predict future streamflow. This study also utilized the quantified flows in the hydraulic model simulation. The hydraulic simulation is used to forecast the extent of the floodplains in the future climate, which is further used for the future flood risk assessment.

The objective of the current study is to forecast the future design discharge and floodplain inundation extent to access the future flood risk imposed upon the Neuse River in NC. The novelty of the study is forecasting the extent of the floodplain and assessment of the risk of the flooding in future climate using the streamflow projection data, that are made available under CMIP6. This study also compared the future and existing FEMA flooding scenarios with the calculated future design discharge to understand the increased severity of flooding in future years. Two design discharges, a 100-year return period and a 500-year return period were used in the 1D hydraulic modeling simulation using HEC-RAS to produce flood inundation maps. Moreover, this study evaluates the difference in the extent of historic flooding with the projected future flooding under the changing climate. The hazard assessment and vulnerability assessment were employed to analyze the anticipated flood risk and help shed light on the severity of risk within the study area. Finally, the risk zone mapping was performed using the projected design discharges. The result obtained from this research might be beneficial for responding to the following questions:

(a)   What would be the impact of climate change on future streamflow, and how will it affect the flood frequency?
(b)   Under the projected design discharge, what would be the future change in flood extent and patterns?
(c)   By what times would the future flood risk increase under the climate change scenarios compared to existing FEMA scenario?

This study forecasted the future floodplain inundation area. It assessed the future flood risk to determine the extent of flood-affected urban and agricultural areas due to increasing streamflow. The outcome of this study will enable policymakers to employ better water resources management measures and lower the risk under the future climate.

2. Study Area and Data Used

2.1. Study Area

The study area was selected in a reach of Neuse River that originates from the confluence Eno and Flat River at Durham County, NC. The Neuse River flows toward the southeast United States in between the piedmont and Pamlico Sound [65,66]. The study reach is 32 km long, and extends from a latitude and longitude of 35.23° N, 77.77° W at the river's upstream with an elevation of 10.73 m to latitude and longitude of 35.25° N, 77.58° W at the river's downstream with an elevation of 4.15 m. The United States geological survey (USGS) gauge station at Neuse River, Kinston (Station ID 02089500) is located 1.2 km upstream from the downstream end of the study reach at the elevation of 3.3 m above NGVD29. The agricultural land, residential area, and wetlands dominate the study reach. The city of Kinston is the principal city within the study area, as shown in Figure 1a,b. This city has experienced numerous flooding events in the past due to extreme rainfall. The study area is a humid subtropical climate with the highest temperature in July and maximum rainfall in September with the average annual precipitation of 1235.96 mm [67]. In the past, this region suffered many extreme flood events due to higher category hurricanes such as hurricane Hazel, Fran, Floyd, and Matthews. Recently, in 2018 hurricane Florence made a devastating impact in the city of Kinston due to storm surge with a record-breaking gage height for that area along with landfall in a different area of NC [68]. The selected reach, Neuse River, along with the city of Kinston, is located in Lenoir County, NC, as shown in Figure 1b. This study used the existing information of the study area from a FEMA flood insurance study (FIS) report [69], hereafter referred to as FFR within this article. Figure 1c shows the digital elevation map (DEM) with the elevation information of the study reach along with the FFR assigned cross-sections.

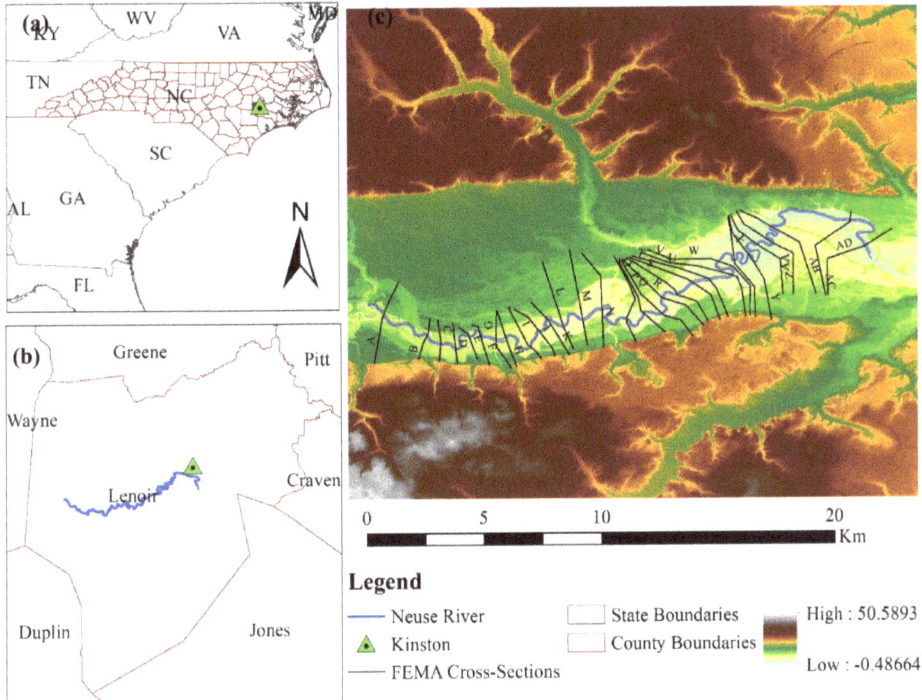

Figure 1. Study area: (**a**) States boundaries outlining counties of North Carolina (NC) including Kinston City, (**b**) Lenoir county showing the Kinston City at downstream of the study reach, (**c**) existing information of the study area from a FEMA flood insurance study (FIS) report (FFR) cross-sections and their labels along with their elevation.

2.2. Dataset

Using single GCM can have more uncertainties in the results [49], so this study utilizes scenarios of CMIP6 consisting of two or more GCMs. Specifically, this study uses four scenarios among twelve available scenarios. Eight scenarios were eliminated due to having only one GCM, and daily streamflow data from selected scenarios are used to evaluate the peak streamflow of the study reach. In CMIP6-AOGCMs, three GCMs for the historic year were available with several climatic projections which were obtained from the model institute named Centre National de Recherches Meteorologiques/Centre Europeen de Recherche et Formation Avancees en Calcul Scientifique (CNRM-CERFACS). A long modeling period from 1950 to 2014 is taken as a historic period. For future streamflow datasets, four different scenarios i.e., SSP1-2.6, SSP2-4.5, SSP3-7.0, and SSP5-8.5, each having two different GCMs from 2036 to 2100 are utilized to access the impact of a hydroclimatic variable in future. These four different future scenarios include climate-changing anthropogenic factors, along with the socio-economic developments [34]. Table 1 presents the scenarios used in this research along with the number of ensemble members. The historical observation data are extracted from the USGS gauge station 02089500, from 1950 to 2014. Also, the gridded streamflow data from CMIP6 has centroid near this USGS gauge station.

Table 1. Global climate models' (GCMs') historical and future scenarios used in the study along with their number of ensemble members and modeling institute [70].

Scenarios	Model Name			Modeling Institute
	CNRM-CM6	**CNRM-ESM2**	**CNRM-CM6-HR**	
Historical	√ (24)	√ (5)	√ (1)	CNRM-CFRFACS
SSP5-8.5	√ (5)	√ (6)		CNRM-CFRFACS
SSP3-7.0	√ (5)	√ (6)		CNRM-CFRFACS
SSP2-4.5	√ (5)	√ (6)		CNRM-CFRFACS
SSP1-2.6	√ (5)	√ (6)		CNRM-CFRFACS

The USGS national map viewer provides the DEM (https://viewer.nationalmap.gov/basic/). For a model with higher accuracy, the literature suggests finer resolution DEM [23,71]. However, due to the limitation of available data, this study used 10-m resolution DEM. Figure 1 shows the DEM of the study reach showing the elevation along with the FFR assigned cross-sections. The land use and land cover data were obtained from the website of Multi-Resolution Land Characteristics Consortium (MRLC) [72] provided the land use and land cover data, which provides the Manning's roughness coefficient of different land use. In this study, the most recent National Land Cover Dataset (NLCD) 2016 data were used. The location of the river cross-sections was selected in and between the cross-section assigned by FFR to aid in the calibration of the hydraulic model. Since the detail of structures like dams, the levee location elevations were not readily available, they were not considered in this study. Furthermore, Manning's values were adopted from FFR for the selected reach length of the Neuse River. FEMA developed a hydraulic analysis flood along with the prediction of different year recurrence flows. Since FEMA performs the flood frequency analysis based on a 100-year and 500-year return period, this study followed the same course, i.e., 100-year (0.1% chance) and 500-year (0.2% chance) of annual occurrence of flooding events. Table 2 shows the extreme streamflow discharges obtained from FFR near Kinston City, NC for the Neuse River.

Table 2. Summary of discharge (m^3/s) at USGS gage site 02089500, downstream of study reach given by FFR [69].

Flooding Source	Location	Drainage Area (Sq. Km)	10% Annual Chance (m^3/s)	2% Annual Chance (m^3/s)	1% Annual Chance (m^3/s)	0.2% Annual Chance (m^3/s)
Neuse River	Approximately 1.2 km upstream of the confluence of Adkin branch	6972.25	639.96	982.59	1146.83	1574.42

3. Methods

The methods section is divided into two sub-sections. First, it explains the statistical analysis to predict future design discharge. Secondly, it depicts a detailed guideline of hydraulic analysis to generate the flood inundation maps and further assessment of potential hazard, vulnerability, and risk. Figure 2 presents the steps involved in this study as the flowchart, which are further discussed below in a sequential manner.

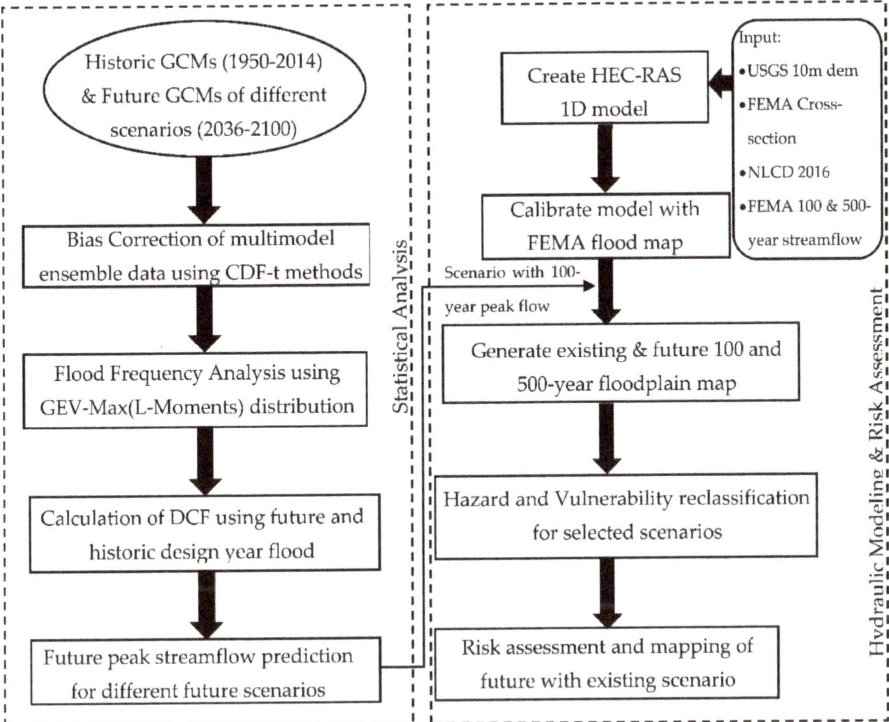

Figure 2. Schematic diagram showing the sequential steps followed to analyze flood frequency, predict the future streamflow, flow routing, floodplain mapping, and risk assessment.

3.1. Statistical Analysis

The following two sub-subsections discuss the statistical analysis involves bias correction of CMIP6 streamflow data and quantification of the future design flows.

3.1.1. Bias Correction

The streamflow data obtained from CMIP6 consist of significant systematic biases, that requires bias correction before further application. Moreover, before bias correction, an ensemble of GCMs was performed for each of the scenarios, so that it would increase the robustness in predicting future change [38,40,73]. In this study, the CDF-t method was chosen for the bias correction of multimodel ensemble CMIP6 streamflow data [50–53]. The CDF-t method develops the relationship between modeled and observed CDF outputs, considering the transformation function "T", where daily observed data were utilized [51]. The descriptive figure to describe the CDF-t methods is as shown in Figure 3, where the modeled future (F_{gf}) and historical climate data (F_{gh}) along with historical observed historical data (F_{sh}) have been used to obtain future bias-corrected data (F_{sf}). Figure 3 shows the F_{sf} as the green dotted line. The bias-corrected value of F_{gh} should fall within the range of F_{sh}. All three arrows, a, b, and c showed the sequential steps of bias correction. Since it is impossible to plot the future change beyond the maximum F_{sh}, arrow "b" moves right toward arrow "c" to intersect the green dot line, which finally generates the bias-corrected data.

The following develops the series of the equation used for the bias correction starting with:

$$T(F_{Gh}(x)) = F_{sh}(x) \tag{1}$$

Let us consider, $u = F_{Gh}(x)$, that gives us $x = F_{Gh}^{-1}(u)$, where $u \in [0,1]$.

Then, Equation (1) is transformed to

$$T(u) = F_{sh}\left(F_{Gh}^{-1}(u)\right) \tag{2}$$

where T represents the functional relationship between the modeled and observed CDF results concerning the historical period.

Validating the Equation (2), the final CDF-t equation is:

$$F_{sf}(x) = F_{sh}\left(F_{Gh}^{-1}\left(F_{Gf}(x)\right)\right). \tag{3}$$

Here, the functional relation is established between observed and modeled streamflow for historical data to be utilized for future periods. Furthermore, the utilization of modeled projection and estimating the CDF of future climate projection is performed, and hence, the bias-corrected data is used for the prediction of future flow.

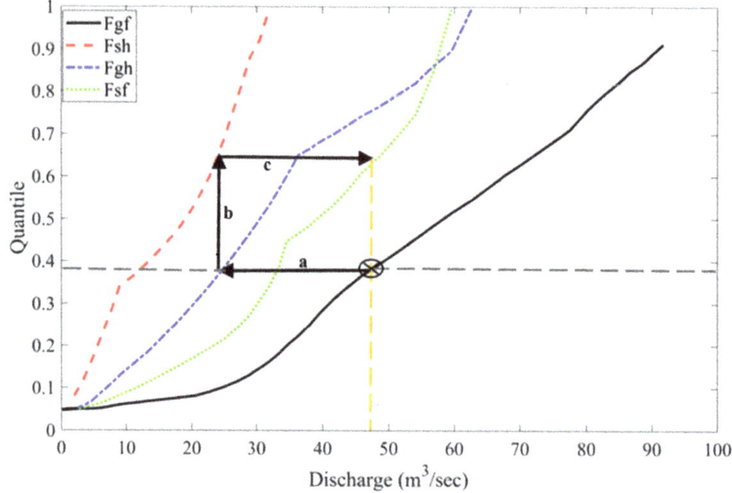

Figure 3. Illustrative figure showing bias correction using CDF-t methods.

3.1.2. Quantification of the Future Design Flow

After the bias correction of streamflow data, the next step is to calculate future design flow. Annual peak streamflow is extracted from each multimodel ensemble scenario (SSP5-8.5, 3-7.0, 2-4.5, 1-2.6) for future time series of 2036 to 2100. Annual peak flow is also calculated for the observed historical time series of 1950 to 2014, from the same USGS gauge station as an earlier step. Then GEV probability distribution is utilized to analyze the annual maximum flow for different recurrence intervals. The equation that was used by the GEV distribution for the annual maxima is given below.

$$\mathrm{GEV}(x : \mu, \sigma, \kappa) = \begin{array}{l} exp\left\{-\left[1 + \kappa\left(\frac{x-\mu}{\sigma}\right)\right]^{-1/\kappa}\right\} \; if \; \kappa \neq 0 \\ exp\left\{-exp\left[-\left(\frac{x-\mu}{\sigma}\right)\right]\right\} \; if \; \kappa = 0 \end{array} \tag{4}$$

In this equation, μ, σ, and κ are GEV parameters respectively representing the location, scale, and shape of the data. Additionally, for shape parameter $\kappa > 0$, $\mu - \sigma/\kappa < x < \infty$; $\kappa = 0$, $-\infty \leq x \leq \infty$; $\kappa < 0$, $-\infty \leq x \leq \mu - \sigma/\kappa$ [58]. Using L-moments, location, scale, and shape parameters are calculated to fit the GEV distribution. Thus, the obtained peak flows for different year return periods are utilized for the estimation of future peak flow. Peak flow for 100 and 500-years is calculated for both climate modeled and observed data with GEV for hydraulic analysis. After the GEV analysis, the delta change method (DCM) is used where the delta change factor (DCF) helps to estimate a different design flood events for evaluation of future peak flows [44]. The main idea of using DCM is to match the currently observed gage streamflow data and FFR streamflow data. For individual scenarios (SSP5-8.5, 3-7.0, 2-4.5, 1-2.6), DCF is calculated and a scenario with higher DCF is chosen for further study. The selected scenario with higher DCF will exhibit an increase in future design flow in its maxima. Thus, GEV generated streamflow for 100-years and 500-year return period are multiplied by the DCF to predict future design streamflow.

3.2. Hydraulic Modeling and Risk Assessment Classification

The hydraulic modeling is performed using the 1D steady model on of HEC-RAS (version 5.0.7). The HEC-RAS model included a total of 95 cross-sections, including 30 existing from FFR to address the critical points along the reach. Manning's roughness coefficient is assigned as suggested by the FFR. The analysis uses the FFR 100-year discharge to calibrate the model with the help of the known water surface elevation levels (WSEL). The model performance is measured using different statistical measures such as Nash-Sutcliffe Efficiency (NSE), Root Mean Square Error (RMSE), Coefficient of Determination (R^2), and Percent-Bias (PBIAS) [74]. After the calibration, the model is routed for the future peak flow obtained for the "quantification of future design flow" step. Hazard and vulnerability assessments are accounted for as the vital steps in accessing the flood risk [27]. For the hazard assessment, different flood characteristics such as water velocity, water depth, and flood extent can be considered as the indicators for hazard classification. This study used water depth for the hazard assessment and attempted to classify the water depth and distinguished the hazard of that area based on the threat posed by flooding on human life. For both extreme events, 100-year, and 500-year, four hazard categories are generated [25,28]. Hazard class is divided into a low hazard (H_1), moderate hazard (H_2), high hazard (H_3), and severe hazard (H_4) class based on the critical flood depth range from 0.8 m to 3.5 m. These hazard classes along with their description are presented in Table 3. In this study, 0.8 m is considered as the level above the ground floor level and 3.5 m is considered as the roof of a single-story building for residential. The human threat is set at the ease of wading at any flooding event.

Table 3. Flood hazard classification and its description considering water depth as an indicator of the degree of hazard.

Hazards Class	Flood Depth (m)	Flood Hazard	Description of Hazard
Low Hazard	<0.8	H_1	Poses less of a hazard to people, and on-foot evacuation can be done.
Moderate Hazard	0.8–1	H_2	On-foot evacuation will be difficult and adult evacuation will be difficult. The infant will be at a serious threat.
High Hazard	1–3.5	H_3	Hazard inside house and evacuation only possible from the roof.
Severe Hazard	>3.5	H_4	All the structures will be underwater, evacuation from the roof will also be a threat as people may be drowned there too.

The Floodplain area in the Neuse River is comprised of different land-use units, which are assigned by NLCD 2016, MRLC. Based on hazard threats on the landforms, risk analysis can be done by utilizing the vulnerability. Hence, as a part of vulnerability assessment, the landforms from NLCD 2016 are reclassified into residential, forest, agriculture, wetlands, and water bodies. The residential area is assigned a value of one (1) and water a value of five (5) representing the flood risk which is lower in residential areas than in areas near the water bodies. Other classified areas such as forest, agricultural land, and wetland were assigned with values of 2, 3, and 4 values, respectively. Table 4 and Figure 4 shows the land use reclassification along with the assigned value. Previous studies showed that the vulnerabilities were assigned based on the projected flooding and socioeconomic conditions [24].

This study uses vulnerabilities based on flood hazard impact and topography for both existing and future scenarios. Furthermore, the future scenarios are comprised of socioeconomic pathways and emission scenarios, that would better depict the threat of expanding floodplains due to climate change. After that, a risk assessment is done by utilizing classified flood hazards due to flooding, which evaluates the threat posed to different landforms based on floodplain depth. Therefore, the magnitude of flood risk is reckoned as a multiplying of the flood hazard and vulnerability, which is inferred physically by assigning values on the equal interval score- a scale from 0 to 20. Scores less than zero are defined as risk free zones, Low Risk Zones are between 0 and 5, Moderate Risk Zones are between 5 and 10, High Risk Zones is between 10 and 15 and Severe Risk Zones are greater than 15 [26]. Then, the risk-map for both the 100 and 500-year design flood for future scenarios is developed. The map portrays the risk that could happen in the future in that area. Moreover, the risk map for the existing FEMA study is also developed to compare the present risk in the study area to the risk projected in the future.

Table 4. Reclassification of the land use data obtained from NLCD (2016) for the study area near Kinston with their assigned value for vulnerability assessment.

Land Classification (NLCD 2016)	Reclassification of Land Use	Score
Developed High Intensity Developed Low Intensity Developed Medium Intensity Developed Open Space	Urbanized Area	1
Deciduous Forest Evergreen Forest Mixed Forest Barren Land Grassland/Herbaceous Shrub/Scrub	Forest	2
Cultivated Crops Pasture/Hay	Agricultural Land	3
Emergent Herbaceous Wetlands Woody Wetlands	Wetlands	4
Open Water	River	5

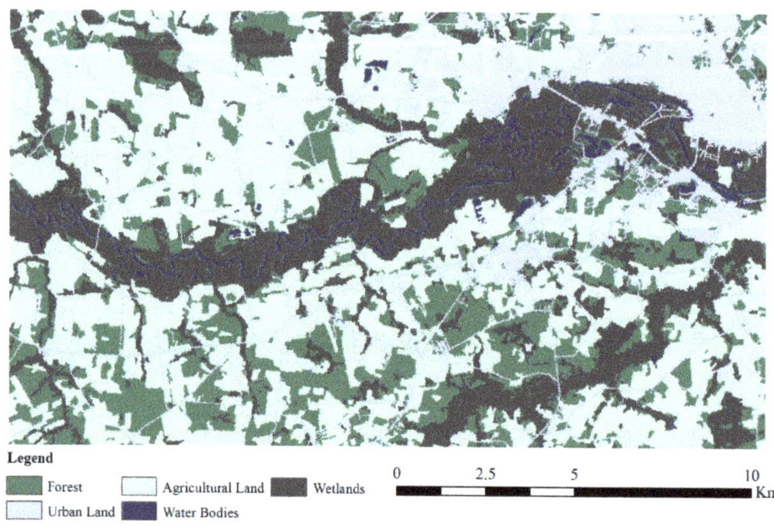

Figure 4. Reclassified NLCD map for the vulnerability assessment.

4. Results

The results are discussed in four ensuing sections. The first section discusses the bias correction results of GCMs obtained from CMIP6. This section also presents future flows that were calculated using bias-corrected data. The second section highlights the simulation results of calibrated hydraulic models that predict the extent of floodplain for different climate scenarios. The third section shows the evaluation of floodplain using classified hazard data. The final section presents the outcomes of risk analysis utilizing reclassified hazard and vulnerability data.

4.1. Flood Frequency Analysis and Performance of Hydraulic Modeling

This study uses daily streamflow data obtained from the climate model projection and USGS gage station, shown in subplots (a) and (b) of Figure 5. The data shows that the range of streamflow is increasing in the Neuse River. A total of 74 streamflow projections from 3 different GCMs of different historical and future climate data were used for the analysis of annual peak flow. The historical data helped to bias-correct future projections for each scenario from the multi-model ensemble GCMs. Figure 5b shows the annual peak flows extracted from the bias-corrected daily streamflow data for each future scenario (i.e., SSP5-8.5, 3-7.0, 2-4.5, 1-2.6). Since the study area is under the risk of catastrophic flooding, the maximum flows are analyzed in this study. Figure 5a,b shows the range of intense flows for future scenarios and historical data. The data in the different scenarios illustrate that the range of flow is greater for SSP3-7.0 and SSP5-8.5, since the projected GHG emissions are at a higher level for these scenarios in comparison to other scenarios.

For the different scenarios representing the different emission pathways, GEV-Max (L-Moments) was utilized to evaluate the design peak flows of different recurrence intervals for different future scenarios (2-year, 5-year, 10-year, 50-year, 100-year, 500-year). Considering the 100-year return period flood, the historic and future scenarios, i.e., SSP1-2.6, SSP2-4.5, SSP5-8.5 were utilized for the calculation of peak discharge, where the discharge was found to be 939.58 m^3/s, 1814.93 m^3/s, 1780.34 m^3/s, 1917.87 m^3/s, and 1921.65 m^3/s, respectively. Across all four future scenarios, SSP5-8.5 generated the maximum flows as the result of higher emission scenarios producing higher radioactive forcing (i.e., 8.5 W m^{-2}) by 2100. For the same scenarios, the 500-year design flow was 1283.11 m^3/s, 4062.07 m^3/s, 3367.09 m^3/s, 3698.73 m^3/s, and 3455.40 m^3/s, respectively. The DCF was calculated by utilizing the different return period discharge from GEV and FFR for the Neuse River. All future return flows were divided by respective discharges according to recurrence intervals from FFR to get DCF values. Figure 5c shows the distribution of DCF for individual scenarios. From Figure 5c, it was inferred that the scenario SSP1-2.6 represents the lowest DCF, i.e., less than one, with the lowest median among all the scenarios. Moreover, scenarios SSP1-2.6, SSP2-4.5, and SSP3-7.0 have DCF values lower than the high emissions scenario SSP5-8.5. The increased value of DCF with increased emission scenarios implies that higher GHG emission, land-use change, and other integrated characteristics are likely to increase the extremes of future streamflow. The Scenario SSP5-8.5 has a higher DCF of 2.045 and 2.69 for 100-year and 500-year design floods. Later in this study, among all four scenarios, SSP5-8.5 based 100-year and 500-year design discharge are used to develop floodplain inundation maps. DCF value implies future design flows for 100-year and 500-year return periods at 2345.52 m^3/s and 4239.88 m^3/s, respectively. When comparing both the 100-year future and existing flow, the future flow is more than two times the existing flow. Similarly, in the case of a 500-year flow, future flow is nearly three times the existing flow. The 100-year future flow even exceeds the existing 500-year flow, demonstrating the higher flood risk in the future.

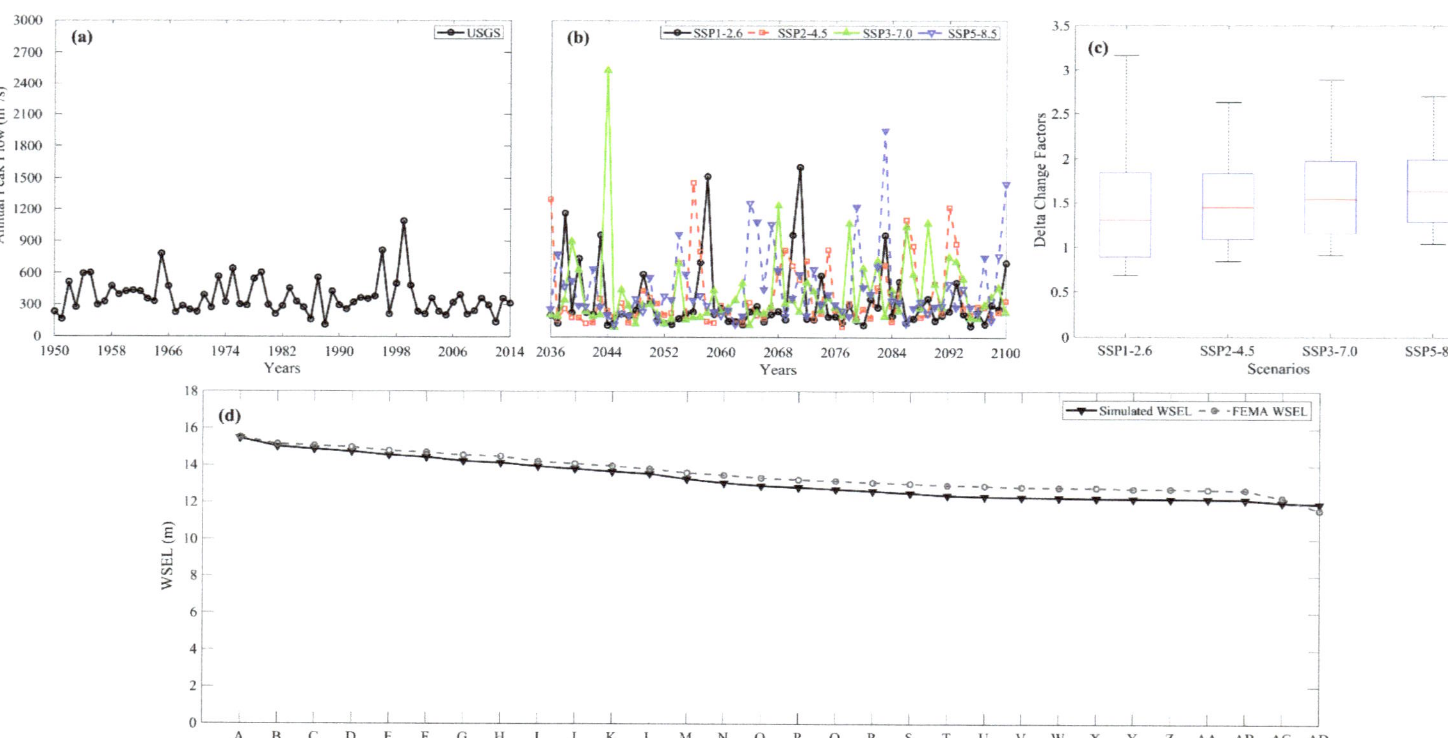

Figure 5. Annual peak streamflow for (**a**) historical and (**b**) future scenarios; (**c**) Box plot for the comparison of different future scenarios DCF using different recurrence intervals; (**d**) Calibration plot of FFR given WSEL versus Simulated WSEL.

The 1D hydraulic model was developed and calibrated using the FFR 100-year design flood. The comparison of observed and simulated WSEL was performed for the calibration using different statistical parameters. Furthermore, the WSEL for all 30 existing cross-sections from the FFR was used in the process. The Manning's "n" used were 0.05–0.06 and 0.12–0.19 for channel and floodplain respectively as suggested by FFR. Figure 5d shows the simulated and observed WSEL for the selected cross-sections. Statistical measures such as NSE, RMSE, R^2, and PBIAS were used to calculate the robustness of the model by comparing the WSEL from the newly developed HEC-RAS model and FFR. The values of NSE, RMSE, R^2, and PBIAS were 0.82, 0.40, 0.98, and −2.64, respectively, and based upon the observed and simulated data. The NSE value closer to 1 suggested that the observed and simulated WSEL were closely fitted. Additionally, the RMSE value of 0.40 shows minimal error regarding observed and simulated WSEL have a close fit. The obtained R^2 value signifies that the observed and simulated WSEL are closely matched with minimal dispersion. The negative value of PBIAS illustrates the overestimation of biases. All the calculated statistical parameters were within an acceptable range and illustrated the robustness of the calibrated hydraulic model, where the predicted WSEL can be utilized to develop a floodplain inundation map.

4.2. Flood Inundation Mapping

The calibrated hydraulic model was used to generate a flood inundation map. The model used DCF implied SSP5-8.5 based future design discharge for 100-year and 500-year return period. The estimated design flows are routed in the HEC-RAS model, and the floodplain inundation areas are mapped using RAS Mapper and ArcGIS. Figure 6 shows the comparison between the floodplain inundation extent generated using the existing FEMA flows and projected future flows. Both, 100-year and 500-year floodplains for future flows are much larger than existing FEMA studies. The floodplain inundation extent generated from the 100-year flood event based on future scenarios was 1.73 times higher than the FFR 100-year floodplain inundation extent. The 500-year floodplain inundation extent was 1.68 times higher than the existing 500-year floodplain extent given by FEMA. The increased floodplain inundation area due to future streamflow that might put Kinston City in a more vulnerable situation. As a coastal state, NC has plenty of low-lying areas used for agricultural purposes [23]. Those areas might be affected due to the future changing streamflow. Additionally, the selected future scenarios, SSP5-8.5, can be used in further studies to address the significant change in streamflow linked with climate change and socio-economic change [34]. It can make a significant difference in the field of future flood studies and emergency flood management efforts.

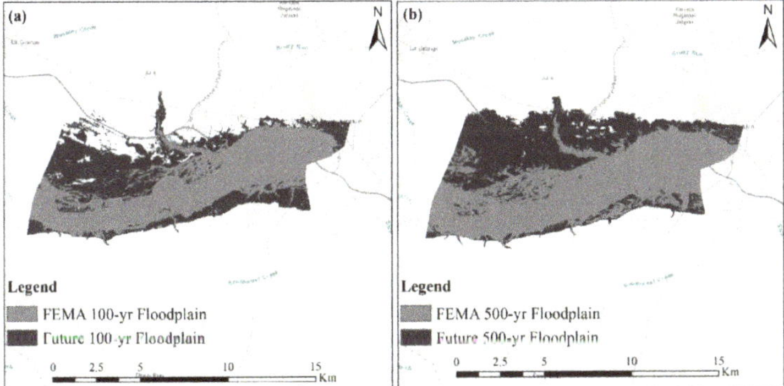

Figure 6. Comparison of flood extent map of Neuse River using the ArcMap (Version 10.7.1) between FEMA and future scenarios for (**a**) 100-year and (**b**) 500-year return period flood events, respectively. Source: Esri, HERE, Garmin, Intermap, increment P Crop., GEBCO, USGS, FAO, NPS, NRCAN, GeoBase, IGN, Kadaster NL, Ordnance Survey, Esri Japan, METI, Esri China (Hong Kong), swisstopo, © OpenStreetMap contributors, and the GIS User Community.

Many flood characteristics were obtained as an output when performing the modeling in HEC-RAS. When generating the floodplain inundation maps for different existing and future scenarios in the Neuse River top width, channel velocity, and flood extent were exported and presented in Figure 7. Figure 7 compares all the variables (top width, channel velocity, and flood extent) for the FEMA and design discharges for the existing 30 cross-sections, acquired from the FIS report, along the reach are shown in Figure 7. The projected future 100-year maximum channel velocity, computed as 1.17 m/s, is higher than the maximum channel velocity. The flood extent area for the future scenario was more than double the area for most of the cross-section in a 100-year flood and more than three times the area for the 500-year projected floodplain when compared to FEMA events. Figure 7a shows the significant changes in the top width for the channel with cross-sections B, C, D, E, F, and H for 100-year flood events. For the 500-year flood, as shown in Figure 7b, cross-sections D, E, and N showed a significant change in top width, which implies that the flood extent is higher in that area. Previous studies [24,26,44] projected 100-year flood events; however, due to this significant change in these flood characteristics in this study, this study was extended up to 500-year flood events. Including the 500-year flood events demonstrates the risk of future flooding, and can improve hydraulic structure designs to mitigate flood risk while considering climate change.

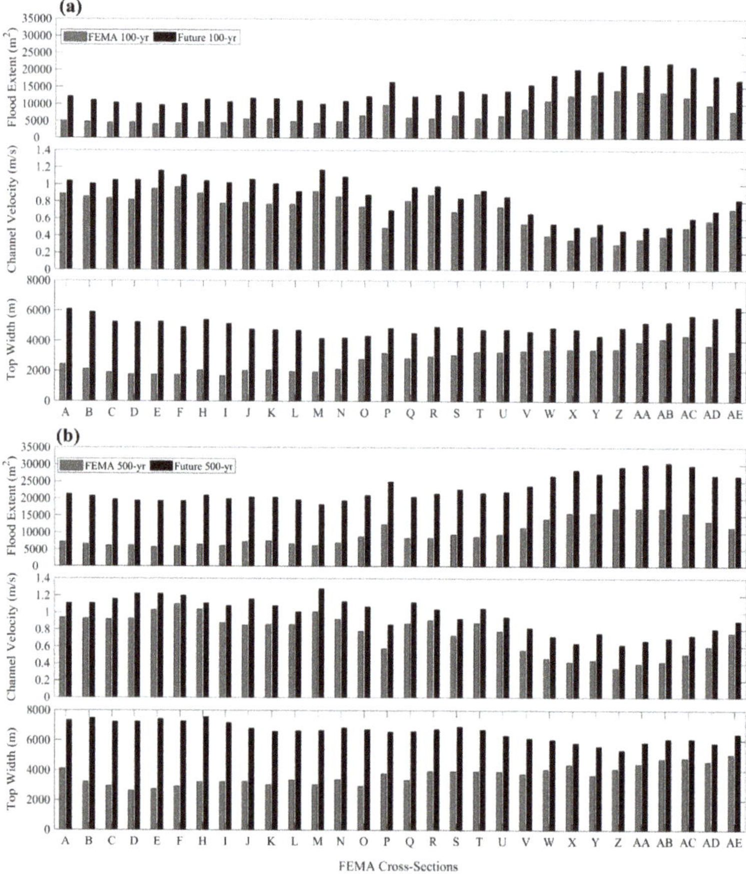

Figure 7. Comparison of flood characteristics such as flood extent, channel velocity, and top width for different flooding scenarios considering (**a**) 100-year and (**b**) 500-year return period flood events, respectively.

4.3. Flood Hazard Assessment

To assess the flood hazard, water depth was used as the quantifiable variable to study the potential threat caused due to existing FEMA and future flooding scenarios for both 100-year and 500-year design floods. All four hazard types were classified for both existing FEMA and future flood scenarios. Figure 8 presents the extent of areas covered by each hazard classification for each scenario. For the 100-year design flood, the future scenario has a higher flood hazard in comparison with the existing FEMA flooding (see Figure 8a,b). Moreover, for this flooding event, the future scenario has a larger, severe hazard floodplain and smaller moderate hazard floodplain. For both FEMA and projected scenarios, the 100-year design flood event has a severe hazard classification with 43.04% (19,219.03 km^2) and 39.72% (30,766.75 km^2), respectively. Since the peak flow was maximum for SSP5-8.5, there could be an increase in the floodplain inundation area with possible hazard classification as low or severe. Figure 8c,d shows the result from the analysis of a 500-year flood hazard assessment. It can be observed that the FEMA flood event has a lesser flood hazard extent than the future scenario (57,976.54 km^2 compared with 97,496.02 km^2). Comparing the existing flood scenario with the projected scenario for 100-year and 500-year flooding events shows the extent of hazard area increased by 1.68 to 1.73 times the existing condition, respectively. The extent of 100-year and 500-year flooding increases from existing to future scenarios, demonstrating there will be potential damage in the future. Thus, these hazard areas combine with a vulnerability factor to identify the degree of risk posed within the study area.

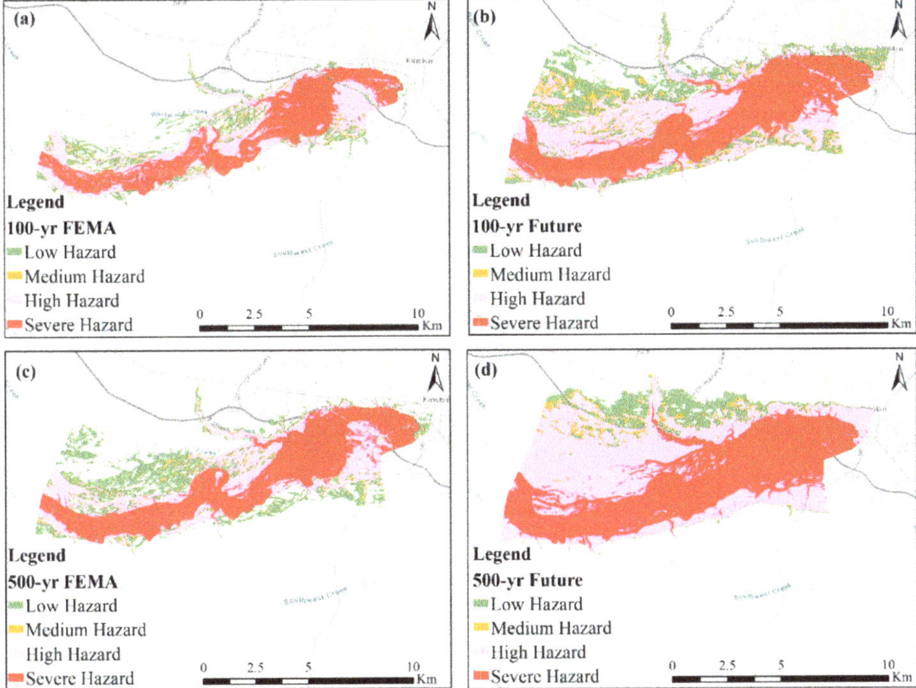

Figure 8. A Comparison of extent of flood hazard utilizing different flooding scenarios, i.e., FEMA and Future, respectively, for (**a,b**) 100 and (**c,d**) 500-year return period flood events. Source: Esri, HERE, Garmin, Intermap, increment P Crop., GEBCO, USGS, FAO, NPS, NRCAN, GeoBase, IGN, Kadaster NL, Ordnance Survey, Esri Japan, METI, Esri China (Hong Kong), swisstopo © OpenStreetMaps contributors, and the GIS User Community.

4.4. Risk Zone Assessment and Mapping

From the intersection of hazard and vulnerability map of FEMA and future climatic scenario SSP5-8.5, risk zone maps are extracted for 100-year and 500-year flooding events. Table 5 summarizes the area covered by each risk zone. The Risk Zone map is comprised of an area under the threat of possible damage [10]. For the FEMA 100-year and 500-year risk maps, a greater extent of the floodplain is covered by a severe risk zone with the area coverage of 18,644.73 km^2 and 23,330.17 km^2. For the 100-year FEMA flood, a moderate risk zone has a lower area coverage among all four risk zone with an area of 6113.23 km^2. Meanwhile in the 500-year flooding event, the flood coverage with high risk zone has a lower area coverage with a total area of 6044.54 km^2. In the future scenario, the 500-year flood moderate risk zone covers an area of 38,869.48 km^2 or 39.75% of the total risk area. For both the future 100-year and 500-year flooding zone, the high risk zone area covered 5211.76 km^2 and 8038.58 km^2, respectively, of the floodplain area. Moreover, Table 5 shows the 100-year and 500-year FEMA flood potential risk area to be 44,659.37 km^2 and 57,979.73 km^2, respectively. Similarly, Table 5 shows the future scenario revealed a moderate risk zone for the 100-year and 500-year return periods at 77,462.86 km^2, and 97,498.36 km^2, respectively. Additionally, it can be inferred that in contrast with the FEMA flood event, the future scenarios have a significant difference in the moderate risk zone. Furthermore, on the evaluation of 100-year and 500-year flood events, the flood risk extent of future scenarios was 1.73 and 1.68 times the existing scenarios, respectively, that shows the increasing flood risk in the future.

Table 5. Flood risk extent based on the zonal risk classification for 100-year and 500-year return period flood event for existing and future scenarios, respectively.

Risk Zone	Existing Scenario (FEMA) (km^2)		Future Scenario (SSP5-8.5) (km^2)	
	100-Year	500-Year	100-Year	500-Year
Low Risk Zone	10,468.76	18,101.99	23,773.57	21,904.91
Moderate Risk Zone	6113.23	10,503.02	22,104.30	38,869.48
High Risk Zone	9432.64	6044.54	5211.76	8038.58
Severe Risk Zone	18,644.73	23,330.17	26,373.23	28,685.39
Total	44,659.37	57,979.73	77,462.86	97,498.36

Figure 9 shows the extent of the risk zone mapped in the study area. In this study, risk zone mapping is an important factor in distinguishing the potential threat for each land use. Figure 9a shows that FEMA 100-year has a minimal threat in urbanized areas and road networks, with a larger area in the risk-free zone. The water bodies, wetlands, and forests were completely in the severe to high risk zone, which will minimize the threat to human life. Some agricultural lands are also observed to be within the low to moderate risk zone, resulting in the potential for lower crop yield during flooding events. Due to the lower elevation in the upstream side of the study area, the risk is exhibited, in contrast to the downstream part. For the 100-year future scenario (Figure 9b), the extent of all risk zone is much larger compared to the FEMA 100-year flood. For this scenario, the risk zone increased in the urbanized area and agricultural lands. Similarly, the wetlands were in a severe risk zone, whereas agricultural lands were in a low to high risk zone. More urbanized areas are located in the low risk zone, which includes residential, industrial, and road networks, which was previously in the no risk zone for the FEMA 100-year flood event. Hence, the increase in low risk extent was higher for future scenarios, which suggests an increase in a potential threat to human settlement in the future. Due to this, there can be an increase in future streamflow that may lead to enlarged flood risk areas. It is necessary to analyze the risk for future and existing 100-year flood events since the events have a higher frequency in history and are likely to happen frequently in the future as well. Moreover, analyzing the future risk would indicate the relative change in socio-economic impact in the study area with the change in the flood hazard area.

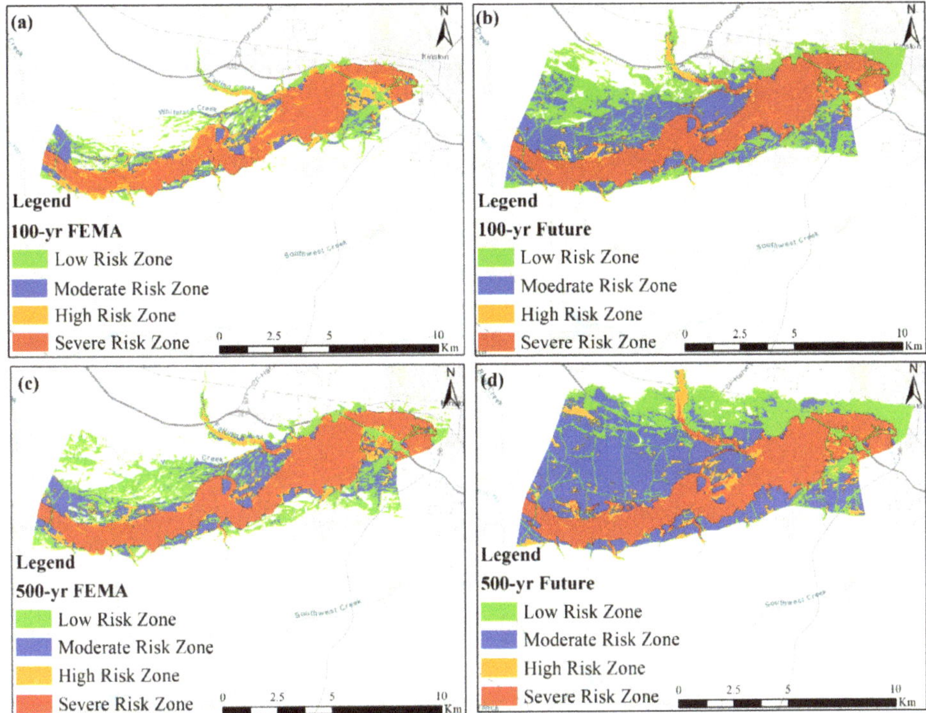

Figure 9. Floodplain mapping based on the risk zone analysis utilizing different flooding scenarios, i.e., FEMA and Future, respectively, for (**a**,**b**) 100-year and (**c**,**d**) 500-year return period flood events. Source: Esri, HERE, Garmin, Intermap, increment P Crop., GEBCO, USGS, FAO, NPS, NRCAN, GeoBase, IGN, Kadaster NL, Ordnance Survey, Esri Japan, METI, Esri China (Hong Kong), swisstopo © OpenStreetMaps contributors, and the GIS User Community.

The risk maps of the 500-year FEMA exhibit all classification of risk zones over the study area. The severe risk zone is larger, extending from waterbodies to the forest and some agricultural lands. The high risk zone covers some parts of the forest, agriculture, and water bodies (Figure 9c). The extent of the low risk zone is greater in an agricultural area near to upstream in the study area and lesser in urban areas, including road networks downstream. As the elevation of the city increases, the risk for the FEMA 500-year flood event decreases in the urbanized area. However, low-lying plain land used for farming purposes is under high risk. The 500-year future scenario results show that most of the urbanized area and agricultural lands are under the low to high risk zone (Figure 9d). Figure 9d also reveals the risk of flood zone with much of that land subjected to a higher risk zone. Future climate scenarios increase the risk that potential floods may have on the urbanized areas of this floodplain. Local water managers can make new development policies to mitigate this risk.

5. Discussion

Flooding can cause colossal damage to human life, settlement, and key infrastructures, resulting in many environmental and socio-economic consequences. The highly urbanized areas located along the floodplain can be at higher risk, even they are highly regulated [26]. Moreover, since ancient times, human populations around the globe have predominately lived near water bodies such as seashores or riverbanks to increase the ease of their living. Around 40% of the world population is currently residing within the 100 km periphery of the coastal areas, which is vulnerable to the different water-related disasters such as sea-level rises and storm surges. The consequences related to these disasters would be more severe due to the highly concentrated population and low elevation in the coastal regions. Furthermore, due to the changes in future streamflow, flood-related hazards are likely to increase for coastal cities like Kinston. Likewise, the change in streamflow is inevitable in the future due to the sea level rises and increased global warming. Accordingly, flood protection planning ensuring minimal loss should be introduced in the future to mitigate flood risk effectively. This study utilized the streamflow projections of CMIP6 to analyze the increase in floodplain inundation due to climate change. Furthermore, different emission scenarios provided by CMIP6 were employed to evaluate the future streamflow, including different forcing level and SSP's. Hence, evaluating the future flow helps us analyze the impact of both climatic and societal change [34], allowing the acknowledgment of a broad range of future streamflow.

For the flood frequency analysis, it is important to use appropriate probability distribution that would be further used to evaluate design floods. For sub-tropical humid climatic conditions, many previous studies suggested the use of GEV as a probability distribution that will likely fit for the prediction of future streamflow [61–63]. Thus, the GEV-Max(L-Moments) was employed for the evaluation of peak flows with different year return period for both existing and future scenarios based on the fact that the climatic condition of a study reach was relevant to that of past research. Within the four different future scenarios, most of them had a DCF value greater than one while considering different return period flood events. These values suggest the future design flow is likely to be higher than the existing flows. The estimated DCF from the multimodel ensemble of four different individual future scenarios is greater than 1, based on the evaluation of the 100-year recurrence interval. The higher DCF generating scenario (SSP5-8.5) was used for this study for the convenience of analysis. For all four scenarios, the design 100-year flow would be nearly double the existing design 100-year flow. The result of the SSP5-8.5 scenario, with the higher DCF among all four scenarios, points to a higher design flow in the future than the existing condition. Moreover, as the DCF increases, the design flow also increases. The SSP5-8.5 scenario is likely to have a higher chance of flooding among all other scenarios, due to a higher predicted design flow. Hence, in further study, future scenarios for both the 100-year and 500-year return period flows were analyzed. DCM was used for the estimation of future flows of different scenarios. Then, the design streamflow from selected future scenarios was routed through the HEC-RAS 1D model for the generation of the floodplain inundation maps to compare it with the existing ones.

Since the DCF is higher for the future scenario SSP5-8.5, the increase in the future design flow is also at its maximum level. So, the maximum design flow from SSP5-8.5 scenario was then employed in the HEC-RAS model for the generation of inundation maps of the study area, Neuse River, NC. The projected future flows were compared with the existing FEMA flows to analyze the impact of climate change in future streamflow. The future 100-year flow was found to be higher than the existing 500-year flow calculated by FEMA. These results suggest an increase in flood extent in the city of Kinston and a higher possibility of a loss of public infrastructures and damage to human settlement. Previous studies suggested an increase in streamflow of the NRB, as the intensity of precipitation was projected to increases due to climatic scenarios [7,13]. For instance, in the last two decades, NC has faced more than three severe hurricanes, resulting in destructive flooding that submerged a portion of the City of Kinston. Previous studies have inferred that the impact of climate change and the change in land use can differ the result of risk analysis, due to change in the extent of floodplain [25,28]. This study uses future flows incorporating climate, land use, and socio-economic changes. Most of the population of Kinston is residing on the bank of the Neuse River, which makes the area more vulnerable to flood hazards. The agricultural land near the riverbank is running through the existing floodplain. Future flooding risks not only affecting the settlement of Kinston city but is also predicted to produce a large impact on local agricultural land, while causing loss of human life and economic harm. Thus, this study performed a zonal risk analysis based on flood hazard and land use, which has shown the severity of risk in land use. It can help local agencies to make a significant effort on the preparedness of those areas depending on the level of risk. Also, the assessment of flood hazards, vulnerability, and risk would lead to greater mitigation of the risk of future flooding.

The current study managed to evaluate the flood hazard area as well as risk zoning so that the risk map can be prepared to depict all four risk zoning areas. Mapping can be a key factor in mitigating flood risk. Policymakers, engineers, and water resource managers can use the risk maps for the planning and constructing facilities that mitigate the risk of future flooding. The risk analysis showed the vulnerability of the community residing in the bank of the Neuse River and the fertile land on the bank of the river. Thus, appropriate research is direly needed for the analysis of future flooding in this area. In this study, different historical and future scenarios hydroclimatic data were employed to predict the future streamflow, which incorporate climatic variability in the future. The future design streamflow was used for the assessment of risk, which can be considered during the planning and building the hydraulic structures so that it can minimize the flood risk incorporated with climate change and socio-economic pathways.

6. Conclusions

This study facilitates the use of different GCMs associated with historical and future scenarios of CMIP6 to estimate the increasing flood risk caused by the changes to streamflow in the future climate. It also shows floodplain changes due to existing and future climatic scenarios. The floodplain area generated using the future climatic scenarios, SSP5-8.5, resulted in nearly two times the floodplain area that is created by existing scenarios for 100-year flood events. This shows the likelihood of an increase in flooding threats under future climate conditions. As manifested by the current change in climate, it is anticipated to increase extreme hydrological events in the future. This will subsequently increase the risk associated with these extremes. The study purposed an approach using different climatic models and observation data that can minimize the adverse impact of change in streamflow in the future. The following points summarize the main conclusion of this study:

1. Bias correction of different scenarios obtained from the multimodel ensemble with the historical data was performed using the CDF-t method. The CDF-t method increases the robustness in evaluating future change in streamflow.

2. For the estimation of the design flow, GEV-Max (L-Moments) was utilized, where SSP5-8.5 was found to have a maximum flow for the 100-year return period.

3. The DCF for most future scenarios were found to be higher than 1, suggesting the increase in future streamflow in comparison with the existing (FEMA) flow.

4. For the 100-year return period flood event, future scenario SSP5-8.5 predicted the maximum increase in the peak flow in Neuse River.

5. HEC-RAS 1D steady modeling was used to simulate the floodplain mapping extent of Neuse River, NC. The result showed a higher extent of flooding for the future 100-year scenario than for the existing FEMA 500-year peak flows.

6. Reclassification and mapping of hazard, vulnerability, and risk were completed utilizing the SSP5-8.5 scenario for the assessment of risk.

7. The extent of different flood risk zone of future flows for 100 and 500-year flood events highlights the increase in potential risk and their severity in the future.

Overall, this research highlights the use of historical and future CMIP6 climate data to forecast the future streamflow for the different return periods. The calculated streamflow is then utilized to develop the future floodplains inundation maps. Using the information from the floodplain maps, a flood risk assessment in terms of a potential threat that can be posed in the study area is performed. As evidenced by the results, the higher GHG emission scenarios are associated with intense future flooding events. These events can pose an adverse effect on the socio-economic factors of the community. On average, flood management structures last for several decades. The structures that are designed using the historic climatic information may not endure future storm events. However, the design of these structures can be optimized by using the forecasted streamflow and by increasing sustainability for the future climatic conditions. The forecasting and risk assessment of such catastrophic events helps policymakers to prepare flood risk mitigation plans and a skeleton for making key decisions in the field of water resource management. As an alternative approach, alteration of land use can be suggested to elevate the management of flood risks sustainably.

Author Contributions: Conceptualization, A.K.; formal analysis, I.P.; M.M.R.; investigation, I.P.; M.M.R.; software, I.P.; M.M.R.; supervision, A.K.; writing—original draft, I.P.; M.M.R.; and A.K.; writing—review and editing, I.P.; M.M.R.; R.T.; and A.K. All authors have read and agreed to the published version of the manuscript.

Funding: This research received no external funding.

Acknowledgments: The authors would like to thank the three reviewers for their valuable comments. The authors are grateful to Bruce DeVantier at Southern Illinois University, Carbondale and Kenneth Lamb at Cal Poly, Pomona for providing assistance with the proofreading of the manuscript. The authors would like to express their appreciation for the support provided by the Office of the Vice-Chancellor for Research at Southern Illinois University, Carbondale. The current study was carried out using various publicly available datasets.

Conflicts of Interest: The authors declare no conflict of interest.

References

1. Allen, R.M.; Dube, O.P.; Solecki, W.; Aragon-Durand, F.; Cramer, W.; Humphreys, S.; Kainuma, M.; Kala, J.; Mahowald, N.; Mulugetta, Y.; et al. Framing and context. In *Global Warming of 1.5 °C. An IPCC Special Report on the Impacts of Global Warming of 1.5 °C above Pre-Industrial Levels and Related Global Greenhouse Gas Emission Pathways, in the Context of Strengthening the Global Response to the Threat of Climate Change, Sustainable Development, and Efforts to Eradicate Poverty*; Masson-Delmotte, V.P., Zhai, H.-O., Portner, D., Roberts, J., Skea, P.R., Shukla, A., Pirani, W., Moufouma-Okia, C., Pean, R., Pidcock, S., et al., Eds.; Intergovernmental Panel on Climate Change: Geneva, Switzerland, 2018; in press.
2. Merz, B.; Aerts, J.; Arnbjerg-Nielsen, K.; Baldi, M.; Becker, A.; Bichet, A.; Blöschl, G.; Bouwer, L.M.; Brauer, A.; Cioffi, F.; et al. Floods and climate: Emerging perspectives for flood risk assessment and management. *NHESS* **2014**, *14*, 1921–1942. [CrossRef]
3. Easterling, D.R.; Meehl, G.A.; Parmesan, C.; Changnon, S.A.; Karl, T.R.; Mearns, L.O. Climate extremes: Observations, modeling, and impacts. *Science* **2000**, *289*, 2068–2074. [CrossRef]
4. Griffin, M.T.; Montz, B.E.; Arrigo, J.S. Evaluating climate change induced water stress: A case study of the lower cape fear basin, NC. *Appl. Geogr.* **2013**, *40*, 115–128. [CrossRef]
5. Middelkoop, H.; Daamen, K.; Gellens, D.; Grabs, W.; Kwadijk, J.C.; Lang, H.; Wilke, K. Impact of climate change on hydrological regimes and water resources management in the Rhine basin. *Clim. Chang.* **2001**, *49*, 105–128. [CrossRef]
6. Roy, L.; Leconte, R.; Brissette, F.P.; Marche, C. The impact of climate change on seasonal floods of a southern Quebec River Basin. *Hydrol. Process.* **2001**, *15*, 3167–3179. [CrossRef]
7. Arnell, N.W.; Gosling, S.N. The impacts of climate change on river flood risk at the global scale. *Clim. Chang.* **2016**, *134*, 387–401. [CrossRef]
8. Hirabayashi, Y.; Kanae, S.; Emori, S.; Oki, T.; Kimoto, M. Global projections of changing risks of floods and droughts in a changing climate. *Hydrol. Sci. J.* **2008**, *53*, 754–772. [CrossRef]
9. De Paola, F.; Giugni, M.; Pugliese, F.; Annis, A.; Nardi, F. GEV parameter estimation and stationary vs. non-stationary analysis of extreme rainfall in African test cities. *Hydrology* **2018**, *5*, 28. [CrossRef]
10. Alfieri, L.; Feyen, L.; Dottori, F.; Bianchi, A. Ensemble flood risk assessment in Europe under high end climate scenarios. *Glob. Environ. Chang.* **2015**, *35*, 199–212. [CrossRef]
11. Bhandari, R.; Kalra, A.; Kumar, S. Analyzing the effect of CMIP5 climate projections on streamflow within the Pajaro River Basin. *Water J.* **2020**, *6*, 5.
12. Chattopadhyay, S.; Jha, M.K. Hydrological response due to projected climate variability in Haw River watershed, North Carolina, USA. *Hydrol. Sci. J.* **2016**, *61*, 495–506. [CrossRef]
13. Johnson, T.; Butcher, J.; Deb, D.; Faizullabhoy, M.; Hummel, P.; Kittle, J.; Sarkar, S. Modeling streamflow and water quality sensitivity to climate change and urban development in 20 US watersheds. *JAWRA J. Am. Water. Resour. Assoc.* **2015**, *51*, 1321–1341. [CrossRef]
14. Arnell, N.W. Climate change and global water resources. *Glob. Environ. Chang.* **1999**, *9*, S31–S49. [CrossRef]

15. Hall, K. *Expected Costs of Damage from Hurricane Winds and Storm-Related Flooding*; Congressional Budget Office: Washington, DC, USA, 2019; pp. 1–48.

16. Guidance of Flood Risk Analysis and Mapping; Hydraulics: One-Dimensional Analysis. Available online: https://www.fema.gov/media-library-data/1484864685338-42d21ccf2d87c2aac95ea1d7ab6798eb/Hydraulics_OneDimensionalAnalyses_Nov_2016.pdf (accessed on 3 January 2020).

17. Brunner, G.W. *HEC-RAS, River Analysis System Hydraulic Reference Manual, Version 5.0*; US Army Corps of Engineers: Davis, CA, USA, 2016; pp. 3–538.

18. Joshi, N.; Lamichhane, G.R.; Rahaman, M.M.; Kalra, A.; Ahmad, S. *Application of HEC-RAS to Study the Sediment Transport Characteristics of Maumee River in Ohio*; World Environmental and Water Resources Congress: Reston, VA, USA, 2019; pp. 257–267.

19. Yang, J.; Townsend, R.D.; Daneshfar, B. Applying the HEC-RAS model and GIS techniques in river network floodplain delineation. *Can. J. Civil. Eng.* **2006**, *33*, 19–28. [CrossRef]

20. Lim, N.J. Performance and Uncertainty Estimation of 1-and 2-Dimensional Flood Models. Master's Thesis, University of Gävle, Gävle, Sweden, June 2011.

21. ShahiriParsa, A.; Noori, M.; Heydari, M.; Rashidi, M. Floodplain zoning simulation by using HEC-RAS and CCHE2D models in the Sungai Maka river. *Sage Open.* **2016**, *9*. [CrossRef]

22. Mehta, D.J.; Ramani, M.M.; Joshi, M.M. Application of 1-D HEC-RAS model in design of channels. *Methodology* **2013**, *1*, 4–62.

23. Peng, A.; Liu, F. Flooding simulation due to hurricane florence in North Carolina with HEC RAS. *arXiv* **2019**, arXiv:1911.09525.

24. Bathi, J.R.; Das, H.S. Vulnerability of coastal communities from storm surge and flood disasters. *Int. J. Environ. Res. Public Health* **2016**, *13*, 239. [CrossRef]

25. Tingsanchali, T.; Karim, F. Flood-hazard assessment and risk-based zoning of a tropical flood plain: Case study of the Yom River, Thailand. *Hydrol. Sci. J.* **2010**, *55*, 145–161. [CrossRef]

26. Mihu-Pintilie, A.; Cîmpianu, C.I.; Stoleriu, C.C.; Pérez, M.N.; Paveluc, L.E. Using high-density LiDAR Data and 2D streamflow hydraulic modeling to improve urban flood hazard maps: A HEC-RAS multi-scenario approach. *Water* **2019**, *11*, 1832. [CrossRef]

27. Klijn, F.; Kreibich, H.; De Moel, H.; Penning-Rowsell, E. Adaptive flood risk management planning based on a comprehensive flood risk conceptualisation. *Mitig. Adapt. Strateg. Glob. Chang.* **2015**, *20*, 845–864. [CrossRef]

28. Tingsanchali, T.; Karim, M.F. Flood hazard and risk analysis in the southwest region of Bangladesh. *Hydrol. Process.* **2005**, *19*, 2055–2069. [CrossRef]

29. Noren, V.; Hedelin, B.; Nyberg, L.; Bishop, K. Flood risk assessment–practices in flood prone Swedish municipalities. *Int. J. Disaster. Risk Reduct.* **2016**, *18*, 206–217. [CrossRef]

30. Eyring, V.; Bony, S.; Meehl, G.A.; Senior, C.A.; Stevens, B.; Stouffer, R.J.; Taylor, K.E. Overview of the Coupled Model Intercomparison Project Phase 6 (CMIP6) experimental design and organization. *Geosci. Model Dev.* **2016**, *9*, 1937–1958. [CrossRef]

31. Meehl, G.A.; Moss, R.; Taylor, K.E.; Eyring, V.; Stouffer, R.J.; Bony, S.; Stevens, B. Climate model intercomparisons: Preparing for the next phase. *Eos Trans. Am. Geophys. Union* **2014**, *95*, 77–78. [CrossRef]

32. Riahi, K.; Van Vuuren, D.P.; Kriegler, E.; Edmonds, J.; O'neill, B.C.; Fujimori, S.; Lutz, W. The shared socioeconomic pathways and their energy, land use, and greenhouse gas emissions implications: An overview. *Glob. Environ. Chang.* **2017**, *42*, 153–168. [CrossRef]

33. Stouffer, R.J.; Eyring, V.; Meehl, G.A.; Bony, S.; Senior, C.; Stevens, B.; Taylor, K.E. CMIP5 scientific gaps and recommendations for CMIP6. *Bull. Am. Meteorol. Soc.* **2017**, *98*, 95–105. [CrossRef]

34. O'Neill, B.C.; Tebaldi, C.; Van Vuuren, D.P.; Eyring, V.; Friedlingstein, P.; Hurtt, G.; Meehl, G.A. The scenario model intercomparison project (ScenarioMIP) for CMIP6. *Geosci. Model Dev.* **2016**, *9*, 3461–3482. [CrossRef]

35. Joshi, N.; Tamaddun, K.; Parajuli, R.; Kalra, A.; Maheshwari, P.; Mastino, L.; Velotta, M. Future changes in water supply and demand for Las Vegas valley: A system dynamic approach based on CMIP3 and CMIP5 climate projections. *Hydrology* **2020**, *7*, 16. [CrossRef]

36. Moradkhani, H.; Baird, R.G.; Wherry, S.A. Assessment of climate change impact on floodplain and hydrologic ecotones. *J. Hydrol.* **2010**, *395*, 264–278. [CrossRef]

37. Alfieri, L.; Bisselink, B.; Dottori, F.; Naumann, G.; de Roo, A.; Salamon, P.; Wyser, K.; Feyen, L. Global projections of river flood risk in a warmer world. *Earths Future* **2017**, *5*, 171–182. [CrossRef]

38. Shrestha, A.; Rahaman, M.M.; Kalra, A.; Jogineedi, R.; Maheshwari, P. Climatological drought forecasting using bias corrected CMIP6 climate data: A case study for India. *Forecasting* **2020**, *2*, 4. [CrossRef]
39. Stevenson, D.S.; Dentener, F.J.; Schultz, M.G.; Ellingsen, K.; Van Noije, T.P.C.; Wild, O.; Bergmann, D.J. Multimodel ensemble simulations of present-day and near-future tropospheric ozone. *J. Geophys. Res. Atmos.* **2006**, *111*. [CrossRef]
40. Nohara, D.; Kitoh, A.; Hosaka, M.; Oki, T. Impact of climate change on river discharge projected by multimodel ensemble. *J. Hydrometeorol.* **2006**, *7*, 1076–1089. [CrossRef]
41. Nam, D.H.; Udo, K.; Mano, A. Future fluvial flood risks in C entral V ietnam assessed using global super-high-resolution climate model output. *J. Flood Risk Manag.* **2015**, *8*, 276–288. [CrossRef]
42. Kay, A.L.; Davies, H.N.; Bell, V.A.; Jones, R.G. Comparison of uncertainty sources for climate change impacts: Flood frequency in England. *Clim. Chang.* **2009**, *92*, 41–63. [CrossRef]
43. Christensen, J.H.; Boberg, F.; Christensen, O.B.; Lucas-Picher, P. On the need for bias correction of regional climate change projections of temperature and precipitation. *Geophys. Res. Lett.* **2008**, *35*. [CrossRef]
44. Nyaupane, N.; Thakur, B.; Kalra, A.; Ahmad, S. Evaluating future flood scenarios using CMIP5 climate projections. *Water* **2018**, *10*, 1866. [CrossRef]
45. Wang, L.; Chen, W. Equiratio cumulative distribution function matching as an improvement to the equidistant approach in bias correction of precipitation. *Atmos. Sci. Lett.* **2014**, *15*, 1–6. [CrossRef]
46. Salvi, K.; Kannan, S.; Ghosh, S. Statistical downscaling and bias-correction for projections of Indian rainfall and temperature in climate change studies. In Proceedings of the 4th International Conference on Environmental and Computer Science, Singapore, 2–4 September 2011; pp. 16–18.
47. Mishra, B.K.; Rafiei Emam, A.; Masago, Y.; Kumar, P.; Regmi, R.K.; Fukushi, K. Assessment of future flood inundations under climate and land use change scenarios in the Ciliwung River Basin, Jakarta. *J. Flood Risk Manag.* **2018**, *11*, S1105–S1115. [CrossRef]
48. Cannon, A.J.; Sobie, S.R.; Murdock, T.Q. Bias correction of GCM precipitation by quantile mapping: How well do methods preserve changes in quantiles and extremes? *J. Clim.* **2015**, *28*, 6938–6959. [CrossRef]
49. Camici, S.; Brocca, L.; Melone, F.; Moramarco, T. Impact of climate change on flood frequency using different climate models and downscaling approaches. *J. Hydrol. Eng.* **2014**, *19*, 04014002. [CrossRef]
50. Guo, L.Y.; Gao, Q.; Jiang, Z.H.; Li, L. Bias correction and projection of surface air temperature in LMDZ multiple simulation over central and eastern China. *Adv. Clim. Chang. Res.* **2018**, *9*, 81–92. [CrossRef]
51. Michelangeli, P.A.; Vrac, M.; Loukos, H. Probabilistic downscaling approaches: Application to wind cumulative distribution functions. *Geophys. Res. Lett.* **2009**, *36*. [CrossRef]
52. Pierce, D.W.; Cayan, D.R.; Maurer, E.P.; Abatzoglou, J.T.; Hegewisch, K.C. Improved bias correction techniques for hydrological simulations of climate change. *J. Hydrometeorol.* **2015**, *16*, 2421–2442. [CrossRef]
53. Famien, A.M.; Janicot, S.; Ochou, A.D.; Vrac, M.; Defrance, D.; Sultan, B.; Noel, T. A bias-corrected CMIP5 dataset for Africa using the CDF-t method: A contribution to agricultural impact studies. *Earth Syst. Dynam.* **2018**, *9*, 313–338. [CrossRef]
54. Yuan, X.; Wood, E.F. Downscaling precipitation or bias-correcting streamflow? Some implications for coupled general circulation model (CGCM)-based ensemble seasonal hydrologic forecast. *Water Resour. Res.* **2012**, *48*. [CrossRef]
55. Hamzah, F.M.; Yusoff, S.H.M.; Jaafar, O. L-moment-based frequency analysis of high-flow at Sungai Langat, Kajang, Selangor, Malaysia. *Sains Malays.* **2019**, *48*, 1357–1366. [CrossRef]
56. Teegavarapu, R.S.; Pathak, C.S. Statistical analysis of precipitation extremes. In *Statistical Analysis of Hydrologic Variables: Methods and Applications*; American Society of Civil Engineering: Reston, VA, USA, 2019; pp. 5–70.
57. Ayuketang Arreyndip, N.; Joseph, E. Generalized extreme value distribution models for the assessment of seasonal wind energy potential of Debuncha, Cameroon. *J. Renew. Energy* **2016**, *2016*, 9. [CrossRef]
58. Hosking, J.R.M.; Wallis, J.R.; Wood, E.F. Estimation of the generalized extreme-value distribution by the method of probability-weighted moments. *Technometrics* **1985**, *27*, 251–261. [CrossRef]
59. Santos, E.B.; Lucio, P.S.; e Silva, C.M.S. Estimating return periods for daily precipitation extreme events over the Brazilian Amazon. *Theor. Appl. Climatol.* **2016**, *126*, 585–595. [CrossRef]
60. Hosking, J.R. L-moments: Analysis and estimation of distributions using linear combinations of order statistics. *J. R. Stat. Soc. Ser. B* **1990**, *52*, 105–124. [CrossRef]
61. Hosking, J.R.M.; Wallis, J.R. L-moments. In *Regional Frequency Analysis: An Approach Based on L-Moments*; Cambridge University Press: Cambridge, UK, 2005; pp. 14–41.

62. Re, M.; Barros, V.R. Extreme rainfalls in se South America. *Clim. Chang.* **2009**, *96*, 119–136. [CrossRef]

63. Shi, P.; Chen, X.; Qu, S.M.; Zhang, Z.C.; Ma, J.L. Regional frequency analysis of low flow based on L moments: Case study in Karst area, Southwest China. *J. Hydrol. Eng.* **2010**, *15*, 370–377.

64. Kasiviswanathan, K.S.; He, J.; Tay, J.H. Flood frequency analysis using multi-objective optimization based interval estimation approach. *J. Hydrol.* **2017**, *545*, 251–262. [CrossRef]

65. NEUSE: River of Peace. Available online: https://www.americanrivers:river/neuse-river/ (accessed on 9 December 2019).

66. Neuse River. Available online: https://www.britannica.com/place/Neuse-River (accessed on 9 December 2019).

67. US Climate Data. Available online: https://www.usclimatedata.com/climate/kinston/north-carolina/united-states/usnc0359 (accessed on 9 December 2019).

68. Stewart, S.R.; Berg, R. *National Hurricane Center Tropical Cyclone Report Hurricane Florence (AL062018)*; National Hurricane Centre: Miami, FL, USA, 2019; pp. 2–98.

69. Flood Insurance Study: A Report of Hazard in Lenoir County, North Carolina and Incorporated Areas. Available online: https://fris.nc.gov/FRIS_WS/PDF/5e19eaf2b15b4aefa17344e46a19c500.pdf (accessed on 25 November 2019).

70. World Research Climate Programme. Available online: https://esgf-node.llnl.gov/search/cmip6/ (accessed on 5 November 2019).

71. Saksena, S.; Merwade, V. Incorporating the effect of DEM resolution and accuracy for improved flood inundation mapping. *J. Hydrol.* **2015**, *530*, 180–194. [CrossRef]

72. Multi-Resolution Land Characteristics (MRLC) Consortium. Available online: https://www.mrlc.gov/ (accessed on 22 November 2019).

73. Sillmann, J.; Kharin, V.V.; Zwiers, F.W.; Zhang, X.; Bronaugh, D. Climate extremes indices in the CMIP5 multimodel ensemble: Part 2. Future climate projections. *J. Geophys. Res. Atmos.* **2013**, *118*, 2473–2493. [CrossRef]

74. Joshi, N.; Bista, A.; Pokhrel, I.; Kalra, A.; Ahmad, S. Rainfall-Runoff Simulation in Cache River Basin, Illinois, Using HEC-HMS. In Proceedings of the World Environmental and Water Resources Congress: Watershed Management, Irrigation and Drainage, and Water Resource Planning and Management, Pittsburgh, PA, USA, 19–23 May 2019; American Society of Civil Engineers: Reston, VA, USA, 2019; pp. 348–360. [CrossRef]

 *forecasting*

Article

Machine Learning-Based Error Modeling to Improve GPM IMERG Precipitation Product over the Brahmaputra River Basin

Md Abul Ehsan Bhuiyan [1,*], Feifei Yang [2], Nishan Kumar Biswas [3], Saiful Haque Rahat [4] and Tahneen Jahan Neelam [5]

[1] Department of Natural Resources and the Environment, University of Connecticut, Storrs, CT 06269-3088, USA
[2] Department of Civil and Environmental Engineering, University of Connecticut, Storrs, CT 06269-3088, USA; feifei.yang@uconn.edu
[3] Department of Civil and Environmental Engineering, University of Washington, Seattle, WA 98195, USA; nbiswas@uw.edu
[4] Department of Chemical and Environmental Engineering, University of Cincinnati, Cincinnati, OH 45220, USA; rahatse@mail.uc.edu
[5] Department of Biological and Environmental Engineering, Cornell University, Ithaca, NY 14853, USA; tn344@cornell.edu
* Correspondence: md.bhuiyan@uconn.edu

Received: 10 June 2020; Accepted: 21 July 2020; Published: 25 July 2020

Abstract: The Integrated Multisatellite Retrievals for Global Precipitation Measurement (GPM) (IMERG) Level 3 estimates rainfall from passive microwave sensors onboard satellites that are associated with several uncertainty sources such as sensor calibration, retrieval errors, and orographic effects. This study aims to provide a comprehensive investigation of multiple machine learning (ML) techniques (Random Forest, and Neural Networks), to stochastically generate an error-corrected improved IMERG precipitation product at a daily time scale and 0.1°-degree spatial resolution over the Brahmaputra river basin. In this study, we used the operational IMERG-Late Run version 06 product along with several meteorological and land surface parameters (elevation, soil type, land type, soil moisture, and daily maximum and minimum temperature) to produce an improved precipitation product in the Brahmaputra basin. We trained, tested, and optimized ML algorithms using 4 years (from 2015 through 2019) of reference rainfall data derived from the rain gauge. The ML generated precipitation product exhibited improved systematic and random error statistics for the study area, which is a strong indication for using the proposed algorithms in retrieving precipitation across the globe. We conclude that the proposed ML-based ensemble framework has the potential to quantify and correct the error sources for improving and promoting the use of satellite-based precipitation estimates for water resources applications.

Keywords: IMERG; SMAP; nonparametric; machine learning; neural network; random forest

---

1. Introduction

The accurate estimation of precipitation incorporates a significant impact on the hydrology, vegetation, natural life, and ecology of any water resource system [1,2]. Despite the fact that precipitation is one of the most imperative parameters for water resource management, documenting precise precipitation information on a global scale is a challenge for climate experts and the scientific community [3–5]. While in situ rain gauge station and weather radar data are the most common sources to obtain precipitation information, satellite-based precipitation products have been recognized as a subordinate source of precipitation data to overcome the restrictions due to inadequate spatial coverage or uneven conveyance from ground-based observations [1,6–9]. However, complex terrain regions with high-altitude satellite precipitation estimates are associated with substantial error due to variability and uncertainty introduced by orographic effects [10–16].

Different satellite-based precipitation products are accessible to supply precipitation at fine spatio-temporal resolutions for a broad range of applications [15,16]. Among these precipitation products, the Integrated Multi-satellite Retrievals for GPM (IMERG) is a combination of features of three multi-satellite precipitation products including (1) Tropical Rainfall Measurement Mission (TRMM) Multi-satellite Precipitation Analysis (TMPA), (2) Climate Prediction Center Morphing (CMORPH) and (3) Precipitation Estimation from Remotely Sensed Information using Artificial Neural Networks (PERSIANN) deemed to have advantages over available satellite-based precipitation products [1,3,17,18]. While IMERG and ground-based observations agree reasonably well for the variations of mean daily precipitation, IMERG tends to overestimate higher monthly precipitation amounts and underestimate dry season precipitation amounts, specifically in multiple regions of Asia [3,19,20].

Moreover, previous studies have assessed the performance of IMERG over varied topographic and geographic features considering seasonal characteristics [19,21–29]. The validation of the IMERG dataset using ground-based observations (i.e., gauges and radar) in conterminous US and Canada suggests an overall improvement in surface precipitation measurement from previous similar products [22,26]. Relative to Tropical Rainfall Measurement Mission (TRMM), IMERG displays improvement associated with the misrepresentation of the rainfall pattern with significant bias reduction in the US conterminous [22] and provides a better correlation in daily and monthly scale in mainland China [27,29,30]. In addition, IMERG showed poor performance even with its improved ability to sense frozen precipitation and deemed to be unreliable over Northern China [31]. Similarly, Murali et al. [24] showed region-specific biases and underestimation for the IMERG products compared to the observed precipitation over the Indian subcontinent. Islam et al. [32] concluded that the performance of IMERG is relatively unsatisfactory for the winter season in Bangladesh and deemed it to be inefficient to estimate the amount of rainfall in general. Such inconsistencies in IMERG products might be associated with various sources of errors causing a detrimental impact on the hydrologic investigation [15,16,33,34], and the need to eliminate these errors for certain regions like Ganges Brahmaputra Meghna (GBM) basin is paramount.

Historically, a significant divergence of opinion on the development of the Brahmaputra river basin has existed among the three demographic giants (namely China, India, and Bangladesh) in this region [35–37]. All these bordering countries are under ever-increasing pressure due to global change, severe water scarcity, and rising demands from population growth [35,38]. Owing to the geopolitical relationships among the countries, Ray et al. [39] identify the absence of an authoritative, dependable, and comprehensive network of basin-wide information on climate as a critical compounding factor in the Brahmaputra basin. A better representative, error-corrected satellite precipitation dataset is essential to fill in the current knowledge gaps in this region.

Therefore, error modeling is vital for improving the use of satellite-based precipitation product (GPM IMERG Precipitation estimates) in precipitation-sensitive applications such as hydrological modeling [40]. The first step to error modeling is to recognize the physical error factors and then evaluate the related error magnitudes. Research on error analysis of the satellite precipitation product has been reported in several past studies, which considered the dependence on precipitation rates and types, as well as surface conditions like soil moisture and land cover [40,41]. A multidimensional satellite rainfall error model (SREM2D) developed by Hossain and Anagnostou [42] has been used in several error modeling studies of satellite rainfall products [43–45]. Bhuiyan et al. [33] recently applied machine learning-based error modeling to evaluate the errors of passive microwave precipitation retrievals based on high-resolution ground radar-rainfall estimates. In that study, they combined meteorological and land surface data from multiple sources to study the impact of land surface conditions (e.g., vegetation cover and soil moisture) on the passive microwave retrieval error.

Recently, nonparametric models have become increasingly popular in weather forecasting, climate change prediction, and the modeling of hydrological processes [14–16,46–50]. Moreover, Non-parametric machine learning techniques such as Quantile Regression Forests (QRF) [51], Random Forests (RF) [52], Classification and Regression Trees (CART) [53], Bayesian Additive Regression Trees (BART) [54], and Neural Networks (NN) [55–57] have become especially popular in hydro-meteorological application [14,33,58–62]. Specifically, NN and RF have been used in several studies to predict precipitation and showed promising results in quantifying precipitation uncertainties for global hydrologic applications [63–67].

The objective of this study is to investigate the use of two machine learning-based error models: Neural Network (NN) and Random forest (RF) in representing the realizations of error-adjusted IMERG rainfall products. The prediction consistency and dependence of NN and RF models in terms of point estimation of GPM IMERG retrieval error are explored carefully to show how the two models perform under different evaluation criteria. The paper is structured as follows: in Section 2, we describe the study area and dataset. Section 3 describes the prediction model, validation methodology, and the performance evaluation error metrics. Evaluation results and discussions are explored in Section 4. Conclusions and recommendations are discussed in Section 5.

2. Study Area and Datasets

2.1. Study Area

The GBM basin consists of the Ganges, Brahmaputra, and Meghna rivers originating from the Himalayas and Vindhya ranges flowing through China, Bhutan, Nepal, India, and Bangladesh and ultimately connecting with the inlet at the Bay of Bengal [68]. In terms of hydrologic vulnerability assessment, GBM deserves special attention for some reasons. For instance, the GBM basin dominates annual flooding cycles, the region of inundation, and the withdrawal of floodwaters based on the hydrologic settings of the adjacent countries surrounded by it [46,69].

Moreover, under changing climate scenarios, dry season rainfall pattern is projected to further decrease for elevated temperature while monsoon rainfall is expected to be more intense resulting from the glacier or early snowmelt. Therefore, a basin level assessment with a more comprehensive evaluation of climate change impacts is mandatory for flood regions such as Nepal, India and Bangladesh [70–72]. Improved IMERG data with limited error might have the utility for the large-scale water resources modeling in GBM to assess climate impact [21,29,39].

Inside the GBM basin, the Brahmaputra River (alternatively known as Yarlung Tsangpo river in China) Basin in Southeast Asia is the fourth largest fluvial system in the world [73]. This basin has a drainage area of about 570,000 km^2 with rugged terrain, accommodating a population of 130 million which is spread over China, India, Bhutan, and Bangladesh [74]. Hydrological modeling for this region is crucial and complex due to its intense seasonal rainfall, unevenly distributed and poorly maintained real-time rain gauge data, and convoluted transboundary issues [75,76].

The study area has a varied topographic gradient from around 8500 m MSL at the origin to about 2 m MSL at the outlet where it meets the Ganges. The upper Brahmaputra river basin lies in the temperate climate zone with mostly unpopulated area whereas the lower Brahmaputra river basin is in a tropical climate that is densely populated and vulnerable to monsoon flooding [77,78]. Hence, this region has a higher number of in situ stations. The study area and the corresponding in situ gauge networks consisting of 120 stations are shown in Figure 1.

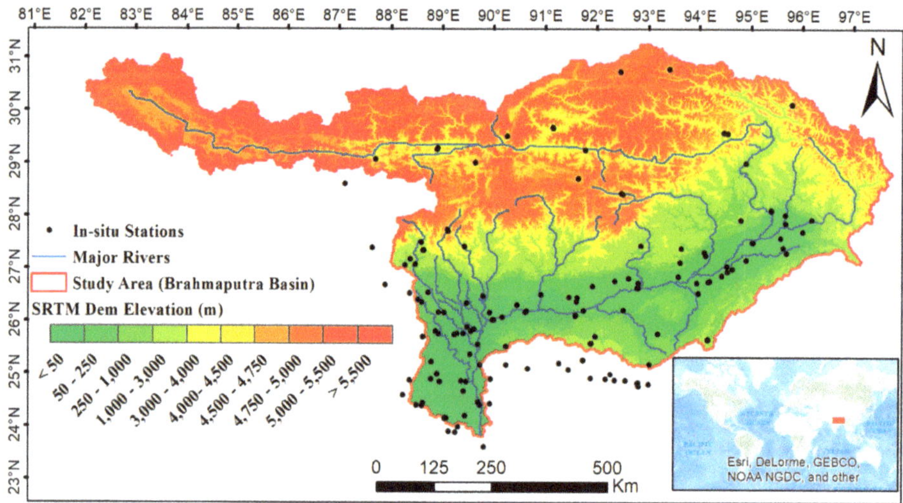

Figure 1. Study Area (Brahmaputra River Basin) with the Shuttle Radar Topographic Mission, SRTM (1 arc second spatial resolution) elevation variation. Solid circles represent gauge locations. The red marked area in the inset map for location reference.

2.2. Datasets

The datasets used in this study span from March 2015 to March 2019. The daily accumulated precipitation datasets from these stations were collected from the Central Water Commission, India; Bangladesh Meteorological Department, Bangladesh, and the Department of Hydrometeorology, Nepal. The gauge measurement were averaged at 0.1 × 0.1 degree grid resolution. For this study, we used the operational IMERG-Late Run version 06 product [79] which was the latest available late run product (https://disc.gsfc.nasa.gov/datasets/GPM_3IMERGDL_06/summary). IMERG precipitation has three different versions, (a) early run; (b) late run and (c) final run. In this study we used late run product which uses a climatological adjustment that incorporates gauge data. The IMERG late run product has both backward and forward morphing and retrieved from different passive microwave (PMW) and infrared (IR) sensors. We used the latest available product (V06) late run version of IMERG because the objective of this study was to focus on the operational use of this precipitation in short-term decision making, cropping, drought management, and water resources planning for the resource-limited stakeholder organizations. Moreover, the latest version of IMERG (V6) has a number of advantages over previous versions such as upgraded GPROF-TMI V05 incorporation into the dataset (https://gpm.nasa.gov/sites/default/files/document_files/IMERG_V06_release_notes_190503.pdf). IMERG (V6) is a high-resolution product available with 0.1 × 0.1 degree spatial and 30 min temporal resolution [6,79]. The dataset's spatial coverage is from 90° N–90° S and temporal coverage is from April 2014 to the present (https://gpm.nasa.gov/data-access/downloads/gpm).

For model forcing, elevation data were extracted from a 30 × 30 DEM (Shuttle Radar Topography Mission) 1 Arc-Second Global (Digital Object Identifier (DOI) number:/10.5066/F7PR7TFT). Soil Moisture Active Passive (SMAP) Level-4 soil moisture data were collected from the National Snow and Ice Data Center for spatial coverage: N: 85.044, S: −85.044, E: 180, W: −180. This dataset has a 9 km Equal-Area Scalable Earth (EASE)-Grid spatial and 3-hourly temporal resolution [80]. The SMAP soil moisture data from March 2015 to March 2019 was used in this study. Daily maximum and minimum land surface temperature was collected from NASA Land Processes Distributed Active Archive Center (LP DAAC). The MOD11C1 version 06 products provide land surface temperature value in a 0.05° by 0.05° Climate Modeling Grid on a daily temporal scale with a latency of approximately 1 day [81]. For land type, USGS Global Land Cover Characterization (GLCC) product with 1 km spatial resolution was used [82]. Soil type was extracted from the FAO Harmonized World Soil Database [83]. Table 1 summarizes land surface and meteorological datasets.

Table 1. Datasets used in this study area.

Data Type	Product	Spatial Resolution	Temporal Resolution	Coverage	Reference/Source
Meteorological Data	Satellite-based Precipitation	0.1° by 0.1°	30 min	Global: 90° N–90° S	https://gpm.nasa.gov/data-access/downloads/gpm
	Soil Moisture	9 km EASE-Grid; Resampled to 0.1° by 0.1°	3 h	Global: 85.044° N–85.044° S	https://nsidc.org/data/SPL4SMGP/versions/4
	Daily Maximum and Minimum Temperature	0.05° by 0.05° Climate Modelling Grid; Resampled to 0.1° by 0.1°	Daily	Global: 90° N–90° S	https://lpdaac.usgs.gov/products/mod11c1v006/
	In-situ Precipitation	Various; Resampled to 0.1° by 0.1°	Daily	Brahmaputra Basin Region	http://cwc.gov.in/ http://live3.bmd.gov.bd/ http://www.dhm.gov.np/
Land Surface Data	SRTM DEM	1 arc second		Global	https://earthexplorer.usgs.gov/
	USGS Land Cover data	1 km grid		Global	https://earthexplorer.usgs.gov/
	FAO Harmonized World Soil Database	30 arc second		Global	http://www.fao.org/soils-portal/soil-survey/soil-maps-and-databases/harmonized-world-soil-database-v12/en/

All the meteorological datasets were resampled to 0.1° by 0.1° using the cubic spline method. The original soil moisture dataset was of 9 km EASE grid with 3 hourly temporal resolution. These datasets were averaged into the daily scale and resampled to 0.1° by 0.1° using the cubic spline method. Daily maximum and minimum temperature data were 0.05° by 0.05°. These data were resampled to the same resolution of the principal forcing data (precipitation) using the cubic spline method. Regarding land surface variables, Shuttle Radar Topographic Mission (SRTM) based Digital Elevation Model (DEM) elevation was extracted for each of the grid locations of the precipitation. Similarly, USGS land cover type and FAO soil type were extracted in each of the grid locations of precipitation. The in-situ precipitation was station-based (in discrete locations). For each day of the study period, all the available in-situ rainfall data were converted into gridded rainfall using Inverse Distance Weightage (IDW) method as mentioned in [84]. Finally, all daily data were mapped to the 0.1° grid chosen to be the final spatial grid for the error model to generate the error-corrected IMERG product. For the error models, the response variable is the rainfall estimate from rain gauge.

3. Methodology

3.1. Precipitation Error Modeling

To develop the error models for the study area we used two machine learning techniques: Random Forests (RF) and Neural Network (NN), to improve GPM IMERG precipitation product. A schematic diagram of the error modeling process is shown in Figure 2. This study devised a randomized and out-of-sample validation experiment to quantify the uncertainty of IMERG precipitation product. Specifically, the two error models, (Neural Network and Random Forest) were developed on the training dataset and were used to predict the holdout dataset, which was applied for testing.

We used "sample" function in R programming to shuffle the row indices of all the dataset to reorder the rows of the dataset randomly. Then, we split 80% of the dataset into the training set and remaining 20% of them into the testing set. Specifically, randomly divided 121,046 rows of data were treated as training and 30,259 rows of the data as testing. To avoid overfitting, we used these independent test data to check the method's accuracy on training data after training which adjusted network structure as well as optimization algorithm parameters of network weight. By using these data and according to the magnitude of mean squared residuals, the network parameters were adjusted. The performance of the error models was evaluated by comparing the error metrics described in Section 3.2.

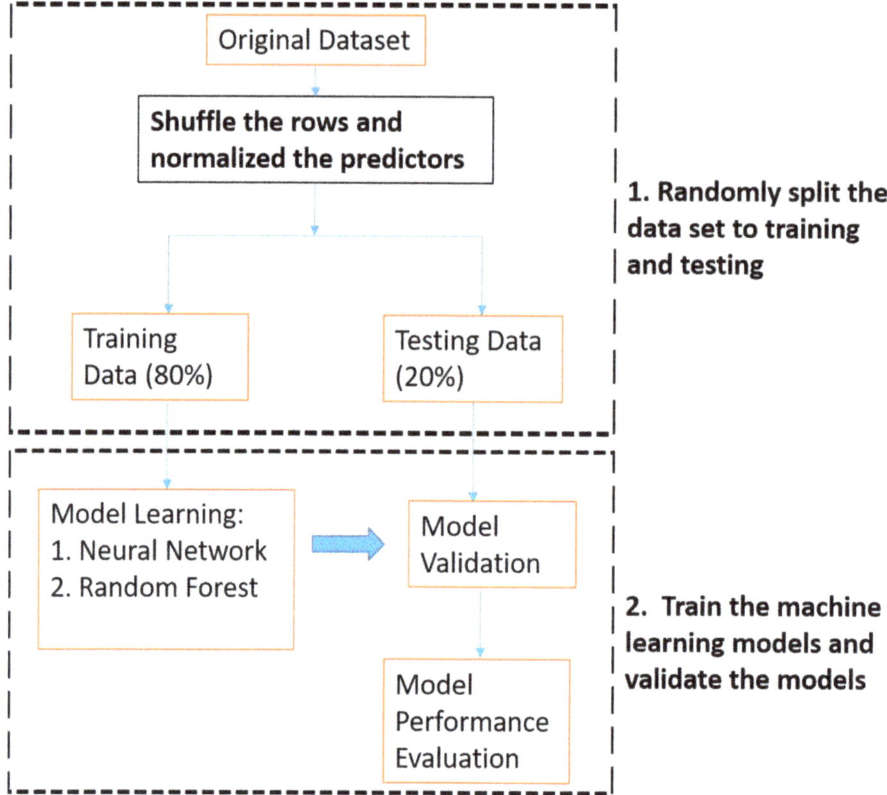

Figure 2. Schematic representation of the prediction process for this study.

3.1.1. Random Forests (RF)

To develop the precipitation error model, we used non-parametric Random Forest (RF) algorithm which uses ensembles ("forests") of classification or regression trees [52]. Each tree depends on the values of a random vector sampled independently and with the same distribution for all trees in the forest. The generalization error for forests converges as the number of trees in the forest becomes large [85,86]. We used R package of "randomForest" and number of trees 1000 for the RF model, and the error for the model was converged before number of trees 1000 as shown in Figure 3.

Initially, to optimize the model, datasets were randomly divided into a training dataset, validation dataset, and test dataset based on the 8:1:1 split rule and the parameters were adjusted for this algorithm. One of the challenges in a data-driven machine learning algorithm is overfitting. In the RF algorithm, each tree chooses and permutes random subsets of input variables at each splitting node, which reduces overfitting and improves the strength of predictions [52]. Therefore, RF utilizes the optimal number "mtry" (size of the random subset of input variables) for split point selection at each node, which introduces randomness in the forests to reduce the correlation between trees [14]. After finalizing the model, we split 80% of the dataset into the training set and remaining 20% of them into the testing set. A schematic diagram is presented in Figure 4. The model testing results are described in Section 4.2.

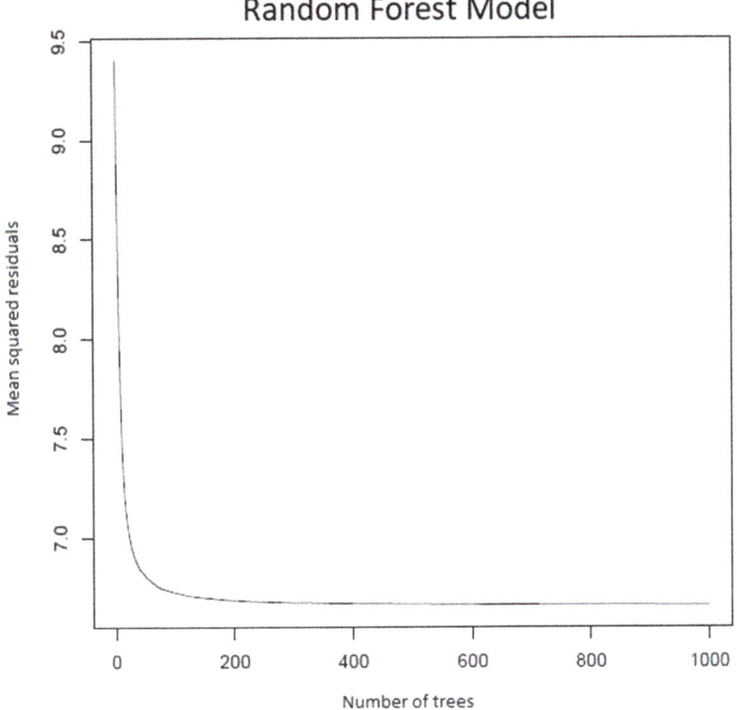

Figure 3. Mean square residual plotted against the number of trees in the random forest.

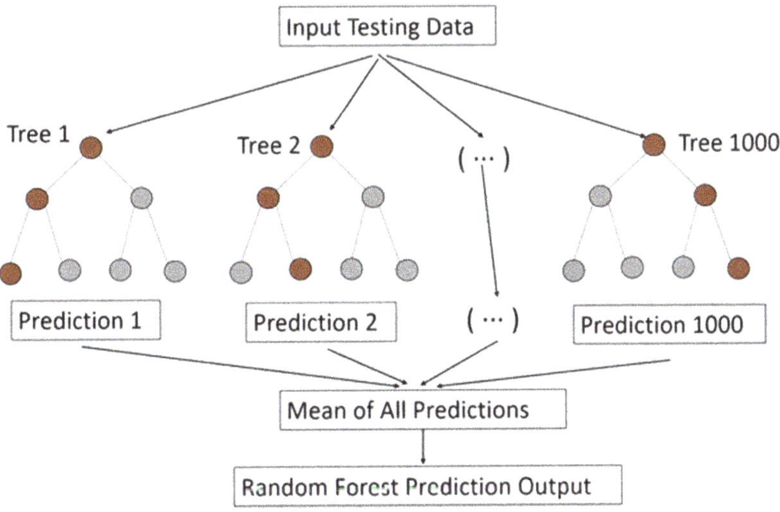

Figure 4. A schematic representation of the random forests (RF).

3.1.2. Neural Network (NN)

The neural network replicates the function of clusters of biological neurons that constitute an animal brain. The fundamental building blocks are called nodes and are used as information processing elements [55–57]. Through a training process, neural networks learn algorithms that can be fitted to the information for detailed data analysis. A schematic representation of Neural Network (NN) is shown in Figure 5. Such learning algorithms are defined by the utilization of a given output that is comparable to the predicted output and by adjusting the parameters as per the comparison.

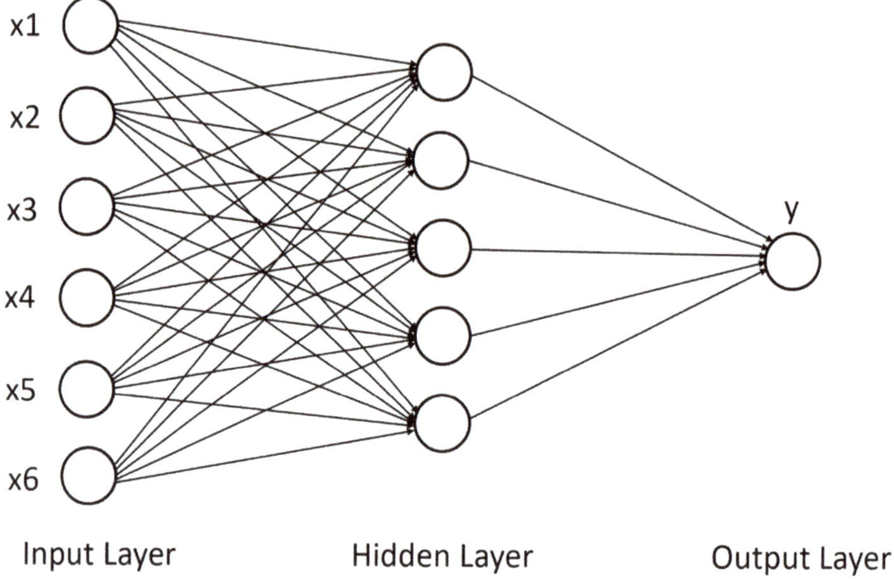

Figure 5. A schematic representation of the neural network (NN).

These predicted outputs are usually transformed through the hidden layers of the neural network from the input data by weights in the parameter [87,88]. When an input enters the node, it gets multiplied by a weight value and the resulting value is either observer or passed to the next layer of the network, which can be helpful to understand the mechanism of such a concept. Suppose a neural network calculates an output $y = f(x)$, for a given input x and weight, w. However, the training process is not completed yet. As such, the predicted output y will be different from the observed output x. To identify such discrepancies, error function E, like the sum of squared errors, SSE $= \sum_{i=1}^{n} (y_i - x_i)^2$ can be used; where y_i is the ith value of the variable to be predicted. All weights keep getting adapted based on the rule of the learning algorithm.

The process stops when all the partial derivative, $\frac{dE}{dw}$ of the error function with respect to the weights are smaller than the defined threshold. Such a methodology was implemented in order to reduce errors for satellite weather data propagation [89]. In this study, NN used backpropagation, namely the resilient backpropagation (RPROP) algorithm [90] which allowed for flexible settings through custom-choice of error and activation function. Finally, to optimize the error model, the calculation of generalized weights [91] was implemented and generated error corrected IMERG prediction. Table 2 summarizes the tuned hyperparameters for the algorithms used in this study.

Table 2. The tuned hyperparameters for RF and NN models.

Random Forest	Neural Network
R Package "randomForest"	R Package "neuralnet"
mtry = 5	Hidden nodes = 5
ntree = 1000	learning rate = 0.01
	stepmax = 10^8
	linear.output = TRUE

3.2. Performance Evaluation Error Metrics

We tasked various error metrics to assess the error model performances. We evaluated the random error component based on the normalized centered root mean square error (*NCRMSE*), and defined as:

$$NCRMSE = \frac{\sqrt{\frac{1}{n}\sum_{i=1}^{n}\left[\hat{y}_i - y_i - \frac{1}{n}\sum_{i=1}^{n}(\hat{y}_i - y_i)\right]^2}}{\frac{1}{n}\sum_{i=1}^{n} y_i} \tag{1}$$

Here, y_i is the reference rainfall, $\hat{y}_i$ is model predicted rainfall, and n is the quantity of samples used in the calculation. *NCRMSE* ranges from 0 (an optimal value) to positive infinity.

To measure the systematic error, we used mean relative error (*MRE*) which is the mean of the relative percentage error, calculated by the normalized average:

$$MRE = \frac{1}{n}\sum_{i=1}^{n}\left(\frac{\hat{y}_i - y_i}{y_i}\right) \tag{2}$$

MRE represents the magnitude and direction of error with positive value referring to overestimation while negative value referring to underestimation.

We applied Theil's 'coefficient of inequality' for the model performances. Theil's inequality coefficient U_1 and U_2 are expressed as below [92]:

$$U_1 = \frac{\sqrt{\frac{1}{n}\sum_{t=1}^{n}(y_t - f_t)^2}}{\sqrt{\frac{1}{n}\sum_{t=1}^{n} y_t^2} + \sqrt{\frac{1}{n}\sum_{t=1}^{n} f_t^2}} \tag{3}$$

$$U_2 = \frac{\sqrt{\sum_{t=1}^{n-1}\left(\frac{f_{t+1} - y_{t+1}}{y_t}\right)^2}}{\sqrt{\sum_{t=1}^{n-1}\left(\frac{y_{t+1} - y_t}{y_t}\right)^2}} \tag{4}$$

Here, the variable of interested is denoted by y_t and the forecast is denoted by f_t. The magnitude of U_1 ranges from 0 and 1 with $U_1 = 0$ suggesting perfect forecast ($y_t = f_t$). Similarly, U_2, a value of zero indicates perfect forecast ($y_{t+1} = f_{t+1}$). U_2 value of 1 indicates how the model performance compares with naïve forecast ($f_{t+1} = y_t$) (Details in [62]).

To assess the relative error metric difference (Δ_{error}) (in %) between model corrected IMERG and original IMERG products we devised the following equation:

$$\Delta_{error} = \frac{\Delta_m - \Delta_i}{\Delta_i} \tag{5}$$

where Δ_m indicates the error metric for model corrected IMERG, and Δ_i represents the error metric for original IMERG product. To calculate the relative reduction of random error, *NCRMSE* error metric is used in Equation (5). Similarly, we used *MRE* in Equation (5) to calculate the relative reduction of systematic error (Details in [15]).

4. Results

4.1. Variable Importance

To construct the error model, the selection of features was based on past research, which demonstrated that several meteorological and land surface features such as satellite-based precipitation, elevation, soil type, land type, soil moisture, and temperature are crucial input features that contribute to the uncertainty of the ML-based error model [14–16]. After choosing these input features, *p*-value experiment and the variable importance methodology [52] were applied to quantify the impact of change from one feature to another.

To assess the impact of the sample size and variability of each variable, *p*-value experiments were examined for this study area. A variable's low *p*-value (<0.05) indicates the rejection of the null hypothesis which means there is a trend in the time series. In this study, *p*-values were determined for all the variables, i.e., IMERG, temperature, soil moisture, elevation, land cover, and soil type. The *p*-values were found close to 0 which are less than the significance level α (alpha) = 0.05 for all the input variables. The result is considered statistically significant by rejecting the null hypothesis. Therefore, the predictor variables are significant in machine learning-based error modeling for this study area.

A variable importance experiment was conducted by calculating the magnitude of the percentage increase in mean square error (%IncMSE) of the model [52,93]. Higher magnitudes of %IncMSE show higher importance of the input features for the error model. The result from the variable importance experiment is displayed in Figure 6. The results showed that all features are comparatively important by producing promising %IncMSE values (0.3–0.8). The level of significance varies marginally for soil moisture, IMERG and temperature (%IncMSE values: 0.65–0.8) among the different variables. Thi sensitivity analysis also demonstrated that other variables are vital by producing decent %IncMSE values (0.3–0.4) for the error modeling.

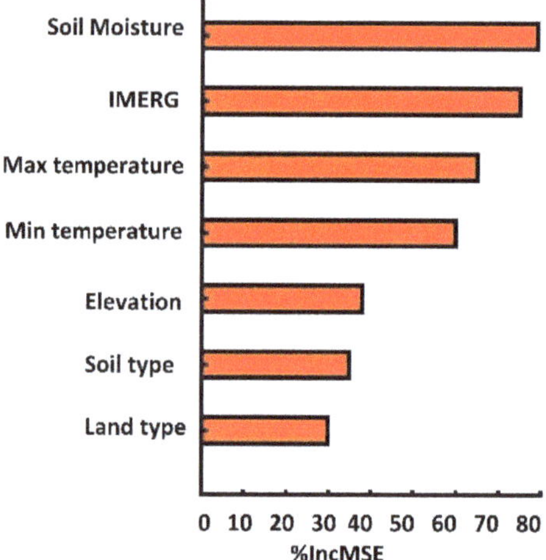

Figure 6. Variable importance plot, where %IncMSE is the percentage increase in mean square error.

4.2. Evaluation of Error Model Corrected Rainfall Rates

In this section, we used Quantile-vs.-Quantile (Q-Q) plot and the error metrics described in Section 3.2 (*NCRMSE, MRE, U_1*, and *U_2*) to compare machine learning-based error model performances. To compare the error-corrected IMERG(V6) precipitation estimates using the two different error models (NN and RF), the Q-Q plots of the original IMERG(V6), error-corrected IMERG(V6), and reference rain rates were produced for the test datasets, as shown in Figure 7. The figure displayed that the NN model corrections exhibited a slight improvement compared to the RF model.

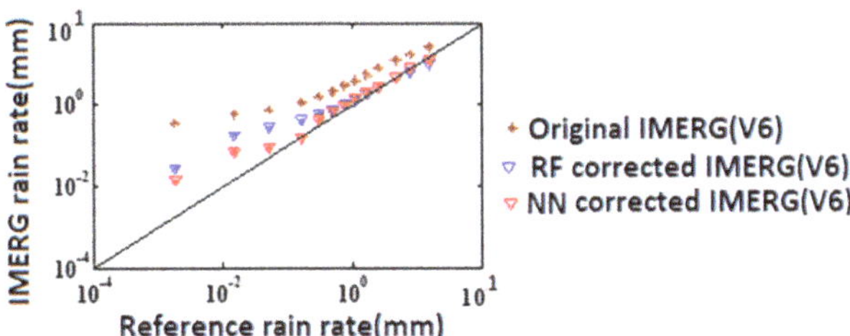

Figure 7. Quantile-vs.-Quantile (Q-Q) diagram of original and model corrected IMERG(V6) rainfall vs. reference rainfall.

The performances of the models were also evaluated in terms of the mean relative error (*MRE*), as shown in Figure 8. *MRE* is calculated for five reference precipitation ranges: rainfall values in the range of <25th, 25th–75th, 75th–90th, 90th–95th, and >95th percentile. The results indicated that NN and RF were able to significantly reduce the systematic error for the >25th percentile. We found low systematic error values for both models corrected IMERG(V6) compared to original IMERG(V6) estimates, which indicated acceptable characterization of estimation uncertainty. The error-corrected IMERG(V6) product exhibited a slightly higher improvement in the NN technique by reducing systematic error compared to the RF model for all five reference precipitation ranges. For the >75th percentile, all individual rainfall datasets (original IMERG(V6), RF-corrected IMERG(V6), NN-corrected IMERG(V6)) showed underestimation. For the low rainfall (<25th), the systematic error (3.2–3.3) slightly reduced for both models, compared to the systematic error (3.7) of the original IMERG(V6). Moreover, the error metric difference (Δ_{error}) considering *MRE* was estimated for the different reference precipitation ranges to show the performances of RF-corrected IMERG(V6), and NN-corrected IMERG(V6), (Table 3). The relative reduction of the systematic error for both model corrected IMERG(V6) is substantial (9–42%) with respect to original IMERG(V6).

The normalized centered root mean square error (*NCRMSE*) metric was examined precisely for the quantification of the performances of the estimates of IMERG. The results are summarized in Figure 8. The results showed that the random error reduced consistently in all products (original IMERG(V6), RF-corrected IMERG(V6), NN-corrected IMERG (V6) as the rainfall rate increased. Both models corrected IMERG(V6) and exhibited lower random error in comparison to the original IMERG(V6) for all precipitation ranges. The NN-corrected IMERG(V6) results exhibited a substantially higher improvement by producing lower random error compared to the RF model. Specifically, for the high rain rates (>95th percentile), the NN error model reported considerably reduced *NCRMSE* values (~0.05) compared to the RF error model. Similarly, for reference rainfall values in the moderate rainfall ranges (>25th percentile to <95th percentile), the results show that the NN-based corrections (*NCRMSE*: 0.15–0.20) bring IMERG estimates closer to the reference precipitation. Furthermore, the error metric difference (Δ_{error}) considering *NCRMSE* for the different models are presented in Table 3. The NN-corrected IMERG(V6) showed substantial relative reduction of random error (37–65%) with respect to the original IMERG(V6) precipitation datasets over the study area. Similarly, RF-corrected IMERG(V6) produced a reasonable relative reduction of random error (~<21%) which also demonstrated the satisfactory performance of RF-corrected IMERG (V6) in comparison to original IMERG (V6). Overall, this study revealed a machine learning-based error model that leads to an advanced error characterization of IMERG precipitation estimation by the significant improvement of random and systematic error.

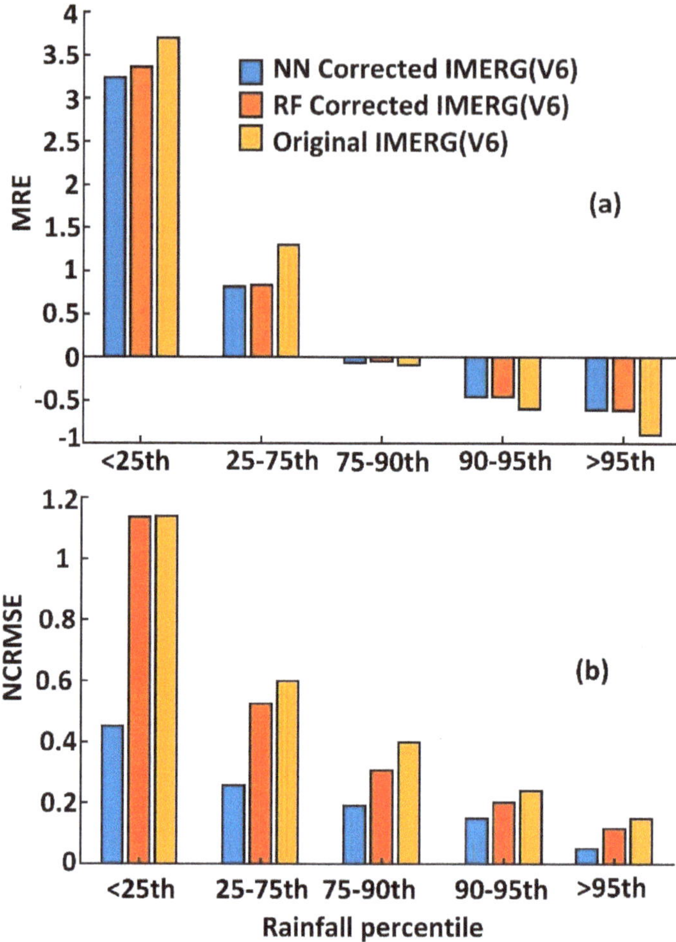

Figure 8. (**a**) Mean relative error of original IMERG(V6) and model corrected IMERG(V6) rain rate; (**b**) Normalized centered root mean square error of original IMERG(V6) and model corrected IMERG(V6) rain rate.

Table 3. Relative reduction of systematic error and random error with respect to original IMERG(V6) rain rate. Results are presented for different precipitation ranges.

Rainfall Percentile	Relative Reduction of Systematic Error		Relative Reduction of Random Error	
	NN	RF	NN	RF
<25th	12%	9%	60%	0%
25–75th	37%	36%	57%	12%
75–90th	24%	42%	52%	23%
90–95th	23%	23%	37%	16%
>95th	32%	32%	65%	21%

In addition, Theil's inequality co-efficient (U_1 and U_2) values for the different models are shown in Figure 9. Theil's inequality co-efficient, U_1 marginally varied 0.15 to 0.17 for the two models, values which are less than the original IMERG(V6) produced U_1 (0.21), as a function of magnitude which showed prominent performances in both models. The magnitude of U_2 greater (lower) than 1 indicates less (more) accurate performance compared to the naïve approach. Moreover, both models also showed similar performances by producing U_2 values close to ~1. Theil's inequality co-efficient, U_2 (1.15) for original IMERG(V6) is greater than both models. These results indicated that the model-corrected IMERG(V6) exhibited a slightly further improvement compared to the original IMERG(V6). Overall, the Theil's inequality co-efficient results showed small relative improvements for the model-corrected IMERG(V6) compared to the original IMERG(V6).

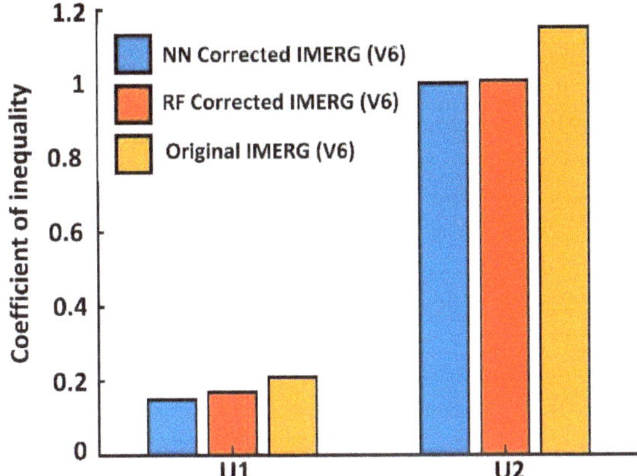

Figure 9. Theil's inequality co-efficient of original IMERG(V6) and model corrected IMERG(V6) rain rate.

4.3. Discussion

Two error models, the random forest (RF) and the neural network (NN), were evaluated based on quantitative error statistics (i.e., *NCRMSE* and *MRE*), and Theil's "coefficient of inequality" statistics (U_1 and U_2). The systematic error for the models varied from overestimation to underestimation as the rain rate increased which is coherent to the findings of [3,14,19,20]. Moreover, the NN model showed promising performance for the moderate and high precipitation rate by displaying a high relative reduction of systematic error (23–37%).

Additionally, the *NCRMSE* metric was also assessed to quantify the effect of precipitation error modeling in reducing the random error component. The study showed that machine learning-based error models significantly reduced random error, and this reduction exhibited rainfall magnitude dependence. Specifically, *NCRMSE* for NN reduced (65%) considerably for the high rain rates (>95th percentile), showing a very high degree of agreement with reference precipitation which was consistent with findings by previous studies [14,15,33]. In addition, the RF and NN techniques considered elevation as a significant feature, which demonstrated the ability of the models to reduce the systematic and random error considerably in the study area. Overall, the performance of the machine learning-based precipitation estimates was consistent with findings by previous studies [14,15,33].

We would like to note that the results shown in Section 4.2 are based on the test dataset and showed improved results by significantly reducing random and systematic errors, which indicates that our model is successfully calibrated and could potentially be useful to predict the independent hydrometeorological dataset. The machine learning-based error models can manipulate the training data in such a way that the actual results expected from the untrained dataset can be quite different from the evaluated results using the training dataset [51,52,94]. Therefore, we considered the representation of extreme (>95th and <25th) precipitation values in the training and testing dataset to make sure that it covered the entire range of the dataset. Applying such a validation approach, the model has good skill on the independent test data in this analysis, which prevents overfitting by producing reliable results.

5. Conclusions

In this study, we investigated machine learning-based precipitation error modeling algorithms to improve the GPM IMERG precipitation product utilizing meteorological and land surface features (elevation, soil type, land type, soil moisture, and daily maximum and minimum temperature) with high-resolution in-situ precipitation rainfall data over the Brahmaputra river basin.

The comparison of NN and RF corrected rainfall values and the reference rainfall values were performed using Q-Q plots and showing satisfactory alignment along the 45-degree line. The error corrected IMERG(V6) results exhibited a slightly higher improvement by NN compared to the RF model. To investigate the accuracy of the error models, validation experiments based on the out-of-sample data approach were used. In terms of systematic and random error metrics, no significant differences were exhibited between two models (RF and NN). Generally, the machine learning-based model is expected not to capture very low and extremely high values successfully [14,93]. This is because the model accuracy is sensitive to sample size and the data representativeness in the training dataset [95,96]. Therefore, very large sample sizes are required for low and extremely high values to quantify the rate of convergence to the underlying cumulative distribution function. Results from quantitative error statistics are consistent in terms of the reduction of the random and systematic error for all the precipitation percentile ranges. This is an indication of how we successfully trained our model instead of overfitting.

The accurate estimation of rainfall in ungauged areas is an essential component to understand water resource systems efficiently. Therefore, extending this machine learning-based error modeling algorithm to the global scale and for other PMW precipitation estimates can be potentially useful. The improvements demonstrated by the error models with independent cross-validation approach indicate the transferability of the error model among complex terrains. Another possible extension of this study is to investigate uses of the PMW ensemble-based error predictions in integrated precipitation algorithms such as NOAA's CMORPH techniques.

Author Contributions: For this research M.A.E.B. developed the precipitation error model, designed, and carried out the analysis of results, and wrote the paper. F.Y. co-designed the analysis of results and contributed to the development of the paper. N.K.B. contributed to the analysis of results and, together with S.H.R., and T.J.N. contributed to the interpretation of results and the writing of the paper. All authors have read and agreed to the published version of the manuscript.

Funding: This research received no external funding.

Acknowledgments: We gratefully acknowledge to GPM (https://gpm.nasa.gov/data-access/downloads/gpm) from where IMERG dataset was obtained for the analysis free of charge. We also acknowledge government organizations: Central Water Commission, India: http://cwc.gov.in/, Bangladesh Meteorological Department, Bangladesh: http://live3.bmd.gov.bd/, Bangladesh Water Development Board, Bangladesh: www.ffwc.gov.bd, Department of Hydrology and Meteorology, Nepal: https://www.dhm.gov.np/, for providing the in-situ data.

Conflicts of Interest: The authors declare no conflict of interest.

References

1. Liu, Z. Comparison of versions 6 and 7 3-hourly TRMM multi-satellite precipitation analysis (TMPA) research products. *Atmos. Res.* **2015**, *163*, 91–101. [CrossRef]
2. Sun, Q.; Miao, C.; Duan, Q.; Ashouri, H.; Sorooshian, S.; Hsu, K. A Review of Global Precipitation Data Sets: Data Sources, Estimation, and Intercomparisons. *Rev. Geophys.* **2018**, *56*, 79–107. [CrossRef]
3. Liu, Z. Comparison of Integrated Multisatellite Retrievals for GPM (IMERG) and TRMM Multisatellite Precipitation Analysis (TMPA) Monthly Precipitation Products: Initial Results. *J. Hydrometeorol.* **2016**, *17*, 777–790. [CrossRef]
4. Steinschneider, S.; Ray, P.; Rahat, S.H.; Kucharski, J. A Weather-Regime-Based Stochastic Weather Generator for Climate Vulnerability Assessments of Water Systems in the Western United States. *Water Resour. Res.* **2019**, *55*, 6923–6945. [CrossRef]

5. Trenberth, K. Changes in precipitation with climate change. *Clim. Res.* **2011**, *47*, 123–138. [CrossRef]

6. Huffman, G.J.; Bolvin, D.T.; Braithwaite, D.; Hsu, K.; Joyce, R.; Kidd, C.; Nelkin, E.J.; Xie, P. NASA Global Precipitation Measurement Integrated Multi-satellitE Retrievals for GPM (IMERG). *Algorithm Theor. Basis Doc.* **2015**, *6*, 30.

7. Hou, A.Y.; Kakar, R.K.; Neeck, S.; Azarbarzin, A.A.; Kummerow, C.D.; Kojima, M.; Oki, R.; Nakamura, K.; Iguchi, T. The Global Precipitation Measurement Mission. *Bull. Am. Meteorol. Soc.* **2014**, *95*, 701–722. [CrossRef]

8. Grecu, M.; Olson, W.S.; Munchak, S.J.; Ringerud, S.; Liao, L.; Haddad, Z.; Kelley, B.L.; McLaughlin, S.F. The GPM Combined Algorithm. *J. Atmos. Ocean. Technol.* **2016**, *33*, 2225–2245. [CrossRef]

9. Smith, E.A.; Asrar, G.; Furuhama, Y.; Ginati, A.; Mugnai, A.; Nakamura, K.; Adler, R.F.; Chou, M.-D.; Desbois, M.; Durning, J.F.; et al. International Global Precipitation Measurement (GPM) Program and Mission: An Overview. In *Measuring Precipitation from Space*; Springer: Dordrecht, The Netherlands, 2007; pp. 611–653.

10. Derin, Y.; Anagnostou, E.; Berne, A.; Borga, M.; Boudevillain, B.; Buytaert, W.; Chang, C.-H.; Delrieu, G.; Hong, Y.; Hsu, Y.C.; et al. Multiregional Satellite Precipitation Products Evaluation over Complex Terrain. *J. Hydrometeorol.* **2016**, *17*, 1817–1836. [CrossRef]

11. Mei, Y.; Anagnostou, E.N.; Nikolopoulos, E.I.; Borga, M. Error Analysis of Satellite Precipitation Products in Mountainous Basins. *J. Hydrometeorol.* **2014**, *15*, 1778–1793. [CrossRef]

12. Houze, R.A., Jr. Orographic effects on precipitating clouds. *Rev. Geophys.* **2012**, *50*, 289. [CrossRef]

13. Roe, G.H. Orographic Precipitation. *Annu. Rev. Earth Planet. Sci.* **2005**, *33*, 645–671. [CrossRef]

14. Bhuiyan, M.A.E.; Nikolopoulos, E.I.; Anagnostou, E.N.; Quintana-Seguí, P.; Barella-Ortiz, A. A nonparametric statistical technique for combining global precipitation datasets: Development and hydrological evaluation over the Iberian Peninsula. *Hydrol. Earth Syst. Sci.* **2018**, *22*, 1371–1389. [CrossRef]

15. Ehsan Bhuiyan, M.A.; Nikolopoulos, E.I.; Anagnostou, E.N. Machine Learning–Based Blending of Satellite and Reanalysis Precipitation Datasets: A Multiregional Tropical Complex Terrain Evaluation. *J. Hydrometeorol.* **2019**, *20*, 2147–2161. [CrossRef]

16. Ehsan Bhuiyan, M.A.; Nikolopoulos, E.I.; Anagnostou, E.N.; Polcher, J.; Albergel, C.; Dutra, E.; Fink, G.; Martínez-de la Torre, A.; Munier, S. Assessment of precipitation error propagation in multi-model global water resource reanalysis. *Hydrol. Earth Syst. Sci.* **2019**, *23*, 1973–1994. [CrossRef]

17. Casse, C.; Gosset, M.; Peugeot, C.; Pedinotti, V.; Boone, A.; Tanimoun, B.A.; Decharme, B. Potential of satellite rainfall products to predict Niger River flood events in Niamey. *Atmos. Res.* **2015**, *163*, 162–176. [CrossRef]

18. Yong, B.; Liu, D.; Gourley, J.J.; Tian, Y.; Huffman, G.J.; Ren, L.; Hong, Y. Global View Of Real-Time Trmm Multisatellite Precipitation Analysis: Implications For Its Successor Global Precipitation Measurement Mission. *Bull. Am. Meteorol. Soc.* **2015**, *96*, 283–296. [CrossRef]

19. Sharifi, E.; Steinacker, R.; Saghafian, B. Assessment of GPM-IMERG and Other Precipitation Products against Gauge Data under Different Topographic and Climatic Conditions in Iran: Preliminary Results. *Remote Sens.* **2016**, *8*, 135. [CrossRef]

20. Tan, M.; Ibrahim, A.; Duan, Z.; Cracknell, A.; Chaplot, V. Evaluation of Six High-Resolution Satellite and Ground-Based Precipitation Products over Malaysia. *Remote Sens.* **2015**, *7*, 1504–1528. [CrossRef]

21. Asong, Z.E.; Razavi, S.; Wheater, H.S.; Wong, J.S. Evaluation of Integrated Multisatellite Retrievals for GPM (IMERG) over Southern Canada against Ground Precipitation Observations: A Preliminary Assessment. *J. Hydrometeorol.* **2017**, *18*, 1033–1050. [CrossRef]

22. Gebregiorgis, A.S.; Kirstetter, P.; Hong, Y.E.; Gourley, J.J.; Huffman, G.J.; Petersen, W.A.; Xue, X.; Schwaller, M.R. To What Extent is the Day 1 GPM IMERG Satellite Precipitation Estimate Improved as Compared to TRMM TMPA-RT? *J. Geophys. Res. Atmos.* **2018**, *123*, 1694–1707. [CrossRef]

23. Kim, K.; Park, J.; Baik, J.; Choi, M. Evaluation of topographical and seasonal feature using GPM IMERG and TRMM 3B42 over Far-East Asia. *Atmos. Res.* **2017**, *187*, 95–105. [CrossRef]

24. Murali Krishna, U.V.; Das, S.K.; Deshpande, S.M.; Doiphode, S.L.; Pandithurai, G. The assessment of Global Precipitation Measurement estimates over the Indian subcontinent. *Earth Space Sci.* **2017**, *4*, 540–553. [CrossRef]

25. Sungmin, O.; Foelsche, U.; Kirchengast, G.; Fuchsberger, J.; Tan, J.; Petersen, W.A. Evaluation of GPM IMERG Early, Late, and Final rainfall estimates using WegenerNet gauge data in southeastern Austria. *Hydrol. Earth Syst. Sci.* **2017**, *21*, 6559–6572.

26. Sunilkumar, K.; Yatagai, A.; Masuda, M. Preliminary Evaluation of GPM-IMERG Rainfall Estimates Over Three Distinct Climate Zones With APHRODITE. *Earth Space Sci.* **2019**, *6*, 1321–1335. [CrossRef]

27. Tang, G.; Ma, Y.; Long, D.; Zhong, L.; Hong, Y. Evaluation of GPM Day-1 IMERG and TMPA Version-7 legacy products over Mainland China at multiple spatiotemporal scales. *J. Hydrol.* **2016**, *533*, 152–167. [CrossRef]

28. Tian, F.; Hou, S.; Yang, L.; Hu, H.; Hou, A. How does the evaluation of GPM IMERG rainfall product depend on gauge density and rainfall intensity? *J. Hydrometeorol.* **2019**, *19*, 339–349. [CrossRef]

29. Xu, R.; Tian, F.; Yang, L.; Hu, H.; Lu, H.; Hou, A. Ground validation of GPM IMERG and TRMM 3B42V7 rainfall products over southern Tibetan Plateau based on a high-density rain gauge network. *J. Geophys. Res. Atmos.* **2017**, *122*, 910–924. [CrossRef]

30. Wang, S.; Liu, J.; Wang, J.; Qiao, X.; Zhang, J. Evaluation of GPM IMERG V05B and TRMM 3B42V7 Precipitation Products over High Mountainous Tributaries in Lhasa with Dense Rain Gauges. *Remote Sens.* **2019**, *11*, 2080. [CrossRef]

31. Chen, F.; Li, X. Evaluation of IMERG and TRMM 3B43 Monthly Precipitation Products over Mainland China. *Remote Sens.* **2016**, *8*, 472. [CrossRef]

32. Islam, M.A. Statistical comparison of satellite-retrieved precipitation products with rain gauge observations over Bangladesh. *Int. J. Remote Sens.* **2018**, *39*, 2906–2936. [CrossRef]

33. Bhuiyan, M.A.E.; Anagnostou, E.N.; Kirstetter, P.-E. A Nonparametric Statistical Technique for Modeling Overland TMI (2A12) Rainfall Retrieval Error. *IEEE Geosci. Remote Sens. Lett.* **2017**, *14*, 1898–1902. [CrossRef]

34. Biemans, H.; Hutjes, R.W.A.; Kabat, P.; Strengers, B.J.; Gerten, D.; Rost, S. Effects of Precipitation Uncertainty on Discharge Calculations for Main River Basins. *J. Hydrometeorol.* **2009**, *10*, 1011–1025. [CrossRef]

35. Jiang, H.; Qiang, M.; Lin, P.; Wen, Q.; Xia, B.; An, N. Framing the Brahmaputra River hydropower development: Different concerns in riparian and international media reporting. *Water Policy* **2017**, *19*, 496–512. [CrossRef]

36. Zawahri, N.A. International rivers and national security: The Euphrates, Ganges-Brahmaputra, Indus, Tigris, and Yarmouk rivers1. *Nat. Resour. Forum* **2008**, *32*, 280–289. [CrossRef]

37. Biba, S. Desecuritization in China's Behavior towards Its Transboundary Rivers: The Mekong River, the Brahmaputra River, and the Irtysh and Ili Rivers. *J. Contemp. China* **2013**, *23*, 21–43. [CrossRef]

38. Feng, Y.; Wang, W.; Liu, J. Dilemmas in and Pathways to Transboundary Water Cooperation between China and India on the Yaluzangbu-Brahmaputra River. *Water* **2019**, *11*, 2096. [CrossRef]

39. Ray, P.A.; Yang, Y.-C.E.; Wi, S.; Khalil, A.; Chatikavanij, V.; Brown, C. Room for improvement: Hydroclimatic challenges to poverty-reducing development of the Brahmaputra River basin. *Environ. Sci. Policy* **2015**, *54*, 64–80. [CrossRef]

40. Oliveira, R.; Maggioni, V.; Vila, D.; Porcacchia, L. Using Satellite Error Modeling to Improve GPM-Level 3 Rainfall Estimates over the Central Amazon Region. *Remote Sens.* **2018**, *10*, 336. [CrossRef]

41. Seyyedi, H.; Anagnostou, E.N.; Kirstetter, P.-E.; Maggioni, V.; Yang, H.; Gourley, J.J. Incorporating Surface Soil Moisture Information in Error Modeling of TRMM Passive Microwave Rainfall. *IEEE Trans. Geosci. Remote Sens.* **2014**, *52*, 6226–6240. [CrossRef]

42. Hossain, F.; Anagnostou, E.N. A two-dimensional satellite rainfall error model. *IEEE Trans. Geosci. Remote Sens.* **2006**, *44*, 1511–1522. [CrossRef]

43. Hossain, F.; Anagnostou, E.N. Assessment of a Multidimensional Satellite Rainfall Error Model for Ensemble Generation of Satellite Rainfall Data. *IEEE Geosci. Remote Sens. Lett.* **2006**, *3*, 419–423. [CrossRef]

44. Maggioni, V.; Reichle, R.H.; Anagnostou, E.N. The Effect of Satellite Rainfall Error Modeling on Soil Moisture Prediction Uncertainty. *J. Hydrometeorol.* **2011**, *12*, 413–428. [CrossRef]

45. Maggioni, V.; Anagnostou, E.N.; Reichle, R.H. The impact of model and rainfall forcing errors on characterizing soil moisture uncertainty in land surface modeling. *Hydrol. Earth Syst. Sci.* **2012**, *16*, 3499–3515. [CrossRef]

46. Schanze, J.; Schwarze, R.; Cartensen, D.; Deilmann, C. Analyzing and managing uncertain futures of large-scale fluvial flood risk systems. Presented at the Managing Flood Risk, Reliability and Vulnerability, Proceedings of 4th International Symposium on Flood Defence, Toronto, ON, Canada, 6–8 May 2008.

47. Croley, T.E., II. Weighted Parametric Operational Hydrology Forecasting. In Proceedings of the World Water & Environmental Resources Congress, Philadelphia, PA, USA, 23–26 June 2003; American Society of Civil Engineers: Reston, VA, USA.

48. Brown, J.D.; Seo, D.-J. A Nonparametric Postprocessor for Bias Correction of Hydrometeorological and Hydrologic Ensemble Forecasts. *J. Hydrometeorol.* **2010**, *11*, 642–665. [CrossRef]
49. Mujumdar, P.P.; Ghosh, S. Climate Change Impact on Hydrology and Water Resources. *ISH J. Hydraul. Eng.* **2008**, *14*, 1–17. [CrossRef]
50. Yenigun, K.; Ecer, R. Overlay mapping trend analysis technique and its application in Euphrates Basin, Turkey. *Meteorol. Appl.* **2012**, *20*, 427–438. [CrossRef]
51. Meinshausen, N. Quantile regression forests. *J. Mach. Learn* **2006**, *7*, 983–999.
52. Breiman, L. Random forests, machine learning 45. *J. Clin. Microbiol.* **2001**, *2*, 199–228.
53. Chipman, H.A.; George, E.I.; McCulloch, R.E. Bayesian CART Model Search. *J. Am. Stat. Assoc.* **1998**, *93*, 935–948. [CrossRef]
54. Chipman, H.A.; George, E.I.; McCulloch, R.E. BART: Bayesian additive regression trees. *Ann. Appl. Stat.* **2010**, *4*, 266–298. [CrossRef]
55. Barnard, E.; Cole, R.A. A neural-net training program based on conjugate-radient optimization. *CSETech* **1989**, 199.
56. Dougherty, M.S.; Cobbett, M.R. Short-term inter-urban traffic forecasts using neural networks. *Int. J. Forecast.* **1997**, *13*, 21–31. [CrossRef]
57. Gunrey, K. *An Introduction to Neural Networks*; UCL Press Limited: London, UK, 1997.
58. Kala, A.; Vaidyanathan, S.G. Prediction of Rainfall Using Artificial Neural Network. In Proceedings of the 2018 International Conference on Inventive Research in Computing Applications (ICIRCA), Coimbatore, India, 11–12 July 2018.
59. Sulaiman, J.; Wahab, S.H. Heavy Rainfall Forecasting Model Using Artificial Neural Network for Flood Prone Area. In *IT Convergence and Security 2017*; Springer: Singapore, 2017; pp. 68–76.
60. Choubin, B.; Zehtabian, G.; Azareh, A.; Rafiei-Sardooi, E.; Sajedi-Hosseini, F.; Kişi, Ö. Precipitation forecasting using classification and regression trees (CART) model: A comparative study of different approaches. *Environ. Earth Sci.* **2018**, *77*, 314. [CrossRef]
61. Kashiwao, T.; Nakayama, K.; Ando, S.; Ikeda, K.; Lee, M.; Bahadori, A. A neural network-based local rainfall prediction system using meteorological data on the Internet: A case study using data from the Japan Meteorological Agency. *Appl. Soft Comput.* **2017**, *56*, 317–330. [CrossRef]
62. Bhuiyan, M.A.E.; Begum, F.; Ilham, S.J.; Khan, R.S. Advanced wind speed prediction using convective weather variables through machine learning application. *Appl. Comput. Geosci.* **2019**, *1*, 100002.
63. Nashwan, M.S.; Shahid, S. Symmetrical uncertainty and random forest for the evaluation of gridded precipitation and temperature data. *Atmos. Res.* **2019**, *230*, 104632. [CrossRef]
64. Herman, G.R.; Schumacher, R.S. Money Doesn't Grow on Trees, but Forecasts Do: Forecasting Extreme Precipitation with Random Forests. *Mon. Weather Rev.* **2018**, *146*, 1571–1600. [CrossRef]
65. Afshin, S.; Fahmi, H.; Alizadeh, A.; Sedghi, H.; Kaveh, F. Long term rainfall forecasting by integrated artificial neural network-fuzzy logic-wavelet model in Karoon basin. *Sci. Res. Essays* **2011**, *6*, 1200–1208.
66. Azadi, S.; Sepaskhah, A.R. Annual precipitation forecast for west, southwest, and south provinces of Iran using artificial neural networks. *Theor. Appl. Climatol.* **2011**, *109*, 175–189. [CrossRef]
67. Sigaroodi, S.K.; Chen, Q.; Ebrahimi, S.; Nazari, A.; Choobin, B. Long-term precipitation forecast for drought relief using atmospheric circulation factors: A study on the Maharloo Basin in Iran. *Hydrol. Earth Syst. Sci.* **2014**, *18*, 1995–2006. [CrossRef]
68. Nishat, B.; Rahman, S.M.M. Water Resources Modeling of the Ganges-Brahmaputra-Meghna River Basins Using Satellite Remote Sensing Data1. *JAWRA J. Am. Water Resour. Assoc.* **2009**, *45*, 1313–1327. [CrossRef]
69. Beran, M.A. Recent advances in statistical flood estimation techniques. In *Flood studies Report—Five Years on*; Thomas Telford Publishing: London, UK, 1981; pp. 25–32.
70. Hossain, F.; Katiyar, N.; Hong, Y.; Wolf, A. The emerging role of satellite rainfall data in improving the hydro-political situation of flood monitoring in the under-developed regions of the world. *Nat. Hazards* **2007**, *43*, 199–210. [CrossRef]
71. Shiklomanov, A.I.; Lammers, R.B.; Vörösmarty, C.J. Widespread decline in hydrological monitoring threatens Pan-Arctic Research. *EosTrans. Am. Geophys. Union* **2002**, *83*, 13. [CrossRef]

72. Sterling, S.M.; Ducharne, A.; Polcher, J. The impact of global land-cover change on the terrestrial water cycle. *Nat. Clim. Chang.* **2012**, *3*, 385–390. [CrossRef]

73. Verma, S.; Mukherjee, A.; Choudhury, R.; Mahanta, C. Brahmaputra river basin groundwater: Solute distribution, chemical evolution and arsenic occurrences in different geomorphic settings. *J. Hydrol. Reg. Stud.* **2015**, *4*, 131–153. [CrossRef]

74. Yang, Y.C.E.; Wi, S.; Ray, P.A.; Brown, C.M.; Khalil, A.F. The future nexus of the Brahmaputra River Basin: Climate, water, energy and food trajectories. *Glob. Environ. Chang.* **2016**, *37*, 16–30. [CrossRef]

75. Bajracharya, S.R.; Palash, W.; Shrestha, M.S.; Khadgi, V.R.; Duo, C.; Das, P.J.; Dorji, C. Systematic Evaluation of Satellite-Based Rainfall Products over the Brahmaputra Basin for Hydrological Applications. *Adv. Meteorol.* **2015**, *2015*, 1–17. [CrossRef]

76. Shrestha, M.S.; Artan, G.A.; Bajracharya, S.R.; Sharma, R.R. Using satellite-based rainfall estimates for streamflow modelling: Bagmati Basin. *J. Flood Risk Manag.* **2008**, *1*, 89–99. [CrossRef]

77. Peel, M.C.; Finlayson, B.L.; McMahon, T.A. Updated world map of the Köppen-Geiger climate classification. *Hydrol. Earth Syst. Sci.* **2007**, *11*, 1633–1644. [CrossRef]

78. Papa, F.; Frappart, F.; Malbeteau, Y.; Shamsudduha, M.; Vuruputur, V.; Sekhar, M.; Ramillien, G.; Prigent, C.; Aires, F.; Pandey, R.K.; et al. Satellite-derived surface and sub-surface water storage in the Ganges–Brahmaputra River Basin. *J. Hydrol. Reg. Stud.* **2015**, *4*, 15–35. [CrossRef]

79. Huffman, G.J.; Stocker, E.F.; Bolvin, D.T.; Nelkin, E.J.; Tan, J. *GPM IMERG Late Precipitation L3 1 Day 0.1 Degree x 0.1 Degree V06*; Andrey, S., Greenbelt, M.D., Eds.; Goddard Earth Sciences Data and Information Services Center (GES DISC), 2019. Available online: https://disc.gsfc.nasa.gov/datasets/GPM_3IMERGDF_06/summary (accessed on 1 May 2020).

80. Reichle, R.; De Lannoy, G.; Koster, R.D.; Crow, W.T.; Kimball, J.S.; Liu, Q. *SMAP L4 Global 3-hourly 9 km EASE-Grid Surface and Root Zone Soil Moisture Geophysical Data*, version 4; NASA National Snow and Ice Data Center Distributed Active Archive Center: Boulder, CO, USA, 2018. [CrossRef]

81. Wan, Z.S.; Hulley, H.G. MOD11C1 MODIS/Terra Land Surface Temperature/Emissivity Daily L3 Global 0.05Deg CMG V006. NASA EOSDIS Land Processes DAAC. last access date: May 26 2020, distributed in netCDF format by the Integrated Climate Data Center (ICDC, icdc.cen.uni-hamburg.de); University of Hamburg: Hamburg, Germany, 2015. [CrossRef]

82. Earth Resources Observation and Science (EROS) Center. Global Land Cover Characterization (GLCC) [Data set]. U.S. Geological Survey, 2017. [CrossRef]

83. FAO/IIASA/ISRIC/ISS-CAS/JRC. *Harmonized World Soil Database (Version 1.1)*; FAO: Rome, Italy; IIASA: Laxenburg, Austria, 2009; Available online: http://www.fao.org/3/a-aq361e.pdf (accessed on 1 May 2020).

84. Biswas, N.K.; Hossain, F. A scalable open-source web-analytic framework to improve satellite-based operational water management in developing countries. *J. Hydroinformatics* **2017**, *20*, 49–68. [CrossRef]

85. Cutler, A.; Cutler, D.R.; Stevens, J.R. Random Forests. In *Ensemble Machine Learning*; Springer: New York, NY, USA, 2012; pp. 157–175.

86. Pal, M. Random forest classifier for remote sensing classification. *Int. J. Remote Sens.* **2005**, *26*, 217–222. [CrossRef]

87. Erb, R.J. Introduction to Backpropagation Neural Network Computation. *Pharm. Res.* **1993**, *10*, 165–170. [CrossRef]

88. Hopfield, J.J. Neural networks and physical systems with emergent collective computational abilities. *Proc. Natl. Acad. Sci.* **1982**, *79*, 2554–2558. [CrossRef]

89. Günther, F.; Fritsch, S. neuralnet: Training of Neural Networks. *R J.* **2010**, *2*, 30. [CrossRef]

90. Riedmiller, M. *Rprop-Description and Implementation Details*; Technical Report; University of Karlsruhe: Karlsruhe, Germany, 1994.

91. Intrator, O.; Intrator, N. Using neural nets for interpretation of nonlinear models. In Proceedings of the Statistical Computing Section, San Francisco, CA, USA, 8–12 August 1993; American Statistical Society: Alexandria, VA, USA, 1993; pp. 244–249.

92. Bliemel, F. Theil's Forecast Accuracy Coefficient: A Clarification. *J. Mark. Res.* **1973**, *10*, 444. [CrossRef]

93. Yang, F.; Watson, P.; Koukoula, M.; Anagnostou, E.N. Enhancing Weather-Related Power Outage Prediction by Event Severity Classification. *IEEE Access* **2020**, *8*, 60029–60042. [CrossRef]

94. Ajiboye, A.R.; Abdullah-Arshah, R.; Hongwu, Q. Evaluating the effect of dataset size on predictive model using supervised learning technique. *Int. J. Comput. Syst. Softw. Eng.* **2015**, *1*, 75–84. [CrossRef]
95. Mondal, A.R.; Bhuiyan, M.A.E.; Yang, F. Advancement of weather-related crash prediction model using nonparametric machine learning algorithms. *SN Appl. Sci.* **2020**, *2*, 1372. [CrossRef]
96. Yang, F.; Wanik, D.W.; Cerrai, D.; Bhuiyan, M.A.E.; Anagnostou, E.N. Quantifying uncertainty in machine learning-based power outage prediction model training: A tool for sustainable storm restoration. *Sustainability* **2020**, *12*, 1525. [CrossRef]

 *forecasting*

Article

Tuning the Bivariate Meta-Gaussian Distribution Conditionally in Quantifying Precipitation Prediction Uncertainty

Limin Wu [1,2]

[1] National Weather Service, Office of Water Prediction, 1325 East-West Highway, Silver Spring, MD 20910, USA; limin.wu@noaa.gov

[2] Lynker, 202 Church Street, SE/Suite 536, Leesburg, VA 20175, USA

Received: 25 November 2019; Accepted: 10 January 2020; Published: 15 January 2020

Abstract: One of the ways to quantify uncertainty of deterministic forecasts is to construct a joint distribution between the forecast variable and the observed variable; then, the uncertainty of the forecast can be represented by the conditional distribution of the observed given the forecast. The joint distribution of two continuous hydrometeorological variables can often be modeled by the bivariate meta-Gaussian distribution (BMGD). The BMGD can be obtained by transforming each of the two variables to a standard normal variable and the dependence between the transformed variables is provided by the Pearson correlation coefficient of these two variables. The BMGD modeling is exact provided that the transformed joint distribution is standard normal. In real-world applications, however, this normality assumption is hardly fulfilled. This is often the case for the modeling problem we consider in this paper: establish the joint distribution of a forecast variable and its corresponding observed variable for precipitation amounts accumulated over a duration of 24 h. In this case, the BMGD can only serve as an approximate model and the dependence parameter can be estimated in a variety of ways. In this paper, the effect of tuning this parameter is studied. Numerical simulations conducted suggest that, while the parameter tuning results in limited improvements in goodness-of-fit (GOF) for the BMGD as a bivariate distribution model, better results may be achieved by tuning the parameter for the one-dimensional conditional distribution of the observed given the forecast greater than a certain large value.

Keywords: meta-Gaussian distribution; Gaussian copula; mallows distance; earth mover's distance; precipitation; precipitation intermittency; uncertainty quantification

---

1. Introduction

The bivariate meta-Gaussian distribution (BMGD), introduced by Kelly and Krzysztofowicz [1], is well suited for modeling the joint distribution of two hydrometeorological variables [2–6]. The BMGD offers a flexible model because each of its marginal variables can be specified independently with any suitable univariate probability distribution. This is a very attractive feature as individual variables in a model may be better fitted by probability distributions from various families. For example, (1) the Weibull and gamma distributions are good candidates for fitting accumulated precipitation amounts [5,7–10], and (2) the lognormal distribution can provide a good fit for mean area average rain rate [11] and for streamflow data in wet seasons [12].

The BMGD was employed in several recent studies in the area of post-processing ensemble weather forecasts. As a way of quantifying predictive uncertainty of precipitation forecasts, Wu et al. [5] explored a mixed type BMGD in producing precipitation ensemble forecasts. Schaake et al. [4] applied the BMGD in producing precipitation and temperature ensemble forecasts. The BMGD is used in producing precipitation and temperature ensemble forecasts based on a number of forecast products from numerical weather prediction models [13]. Validation results indicate that the BMGD is a suitable model for use in producing reliable precipitation and temperature ensemble forecasts for hydrologic ensemble forecasting [14,15]. The BMGD is extended in [16] to a wider class of probability distributions with given marginals, known as meta-elliptical distributions. This extended model was shown by [17] to be flexible in modeling trivariate hydrologic data.

The BMGD can be formulated in two different ways. Although both lead to the same end result in terms of the form of probability distribution, there is a subtle difference in the assumption for the model's dependence structure. In the conventional approach, which is characterized by the construction of a bivariate probability density weighting function [2], one transforms each of the variables to a standard normal variable. A key assumption of this formulation is that the transformed variables are jointly normal. In real-world applications, this assumption may not hold. Herr and Krzysztofowicz [8] conducted a study on the normality hypothesis with the use of 24-h precipitation data collected at a number of rain gauge locations in two river basins in the Appalachian Mountains. Their findings are mixed: for one basin, the normality hypothesis was not rejected in most of the cases; for the other basin, the hypothesis was rejected in most of the cases, albeit possibly of little practical concern. To further illustrate this point, Section 4.2 of this paper gives two examples showing that this normality assumption can be invalid for the cases under consideration. In the next section, it is shown that this assumption can be dropped if we formulate the BMGD using the Gaussian copula to yield an approximate model. With this approach, the BMGD becomes more flexible in that the dependence parameter of the distribution is not constrained by the data for modeling the marginal distributions. The theoretical implication of this relaxation is that the model can be tuned by adjusting the dependence parameter.

This work is primarily concerned with tuning the dependence parameter of the BMGD so that the model can fit the data better. The tuning is investigated for the BMGD as a bivariate distribution model (unconditional tuning) and for the one-dimensional conditional distribution of the observed given the forecast greater than a certain large value (conditional tuning). In the context of generating ensemble forecasts from the BMGD model, the adjustment pertains to a major objective of ensemble post-processing: reliability, which refers to statistical consistency between a priori predicted probabilities and a posteriori observed frequencies of the occurrence of an event under consideration [18]. If the data sample is sufficiently large, then a better fit of the model would translate to more reliable ensemble forecasts. The fit of the model may be improved by selecting more suitable marginal distributions as well as tuning the dependence parameter of the model. In this study, the focus is on model tuning in which both unconditional tuning and conditional tuning are explored.

This paper is organized as follows: Section 2 describes the two different approaches for formulating the BMGD. First, a brief description of the conventional approach is given. Then, a formulation with the copula approach is laid out. Methods used in this study for estimating the dependence parameter is also described. Section 3 describes a bivariate distribution model for an observed variable and its corresponding forecast variable for precipitation amounts. The model accounts for precipitation intermittency explicitly. Equations characterizing the intermittency are derived using calculus and elementary probability theory. In this model, the portion of the joint distribution where both variables are positive is separated out so that it can be modeled by the BMGD. Section 4 describes the numerical experiments conducted in this study. It includes a description of the experimental data and simulation processes, and a discussion of some results. Section 5 provides the discussion, and the final section gives the conclusions.

2. Bivariate Meta-Gaussian Distribution

2.1. Formulation

The conventional approach starts with a density weighting function [1,2]. A normality assumption is made with this approach. The following gives a description of how the BMGD can be obtained, without reference to the density weighting function, for easy comparison with the alternative approach. Consider the joint cumulative distribution function (CDF) of two continuous variables X and Y:

$$F(x,y) := P(X \le x, Y \le y). \tag{1}$$

Denote the CDF of X and Y by $F_X(x)$ and $F_Y(y)$, respectively. Assume that $F_X(x)$ and $F_Y(y)$ are strictly increasing. Let Φ denote the standard normal distribution function and Φ^{-1} its inverse. The X and Y can be transformed to obtain two standard normal variables Z and W: $Z = \Phi^{-1}(F_X(X))$ and $W = \Phi^{-1}(F_Y(Y))$. The transformation involved is known as normal quantile transform (NQT). It follows that $P(X \le x, Y \le y) = P(Z \le z, W \le w)$, where $z = \Phi^{-1}(F_X(x))$ and $w = \Phi^{-1}(F_Y(y))$. Let ρ denote the Pearson product-moment correlation coefficient between Z and W. If we assume that

$$P(Z \le z, W \le w) = \Phi_\rho(z,w), \tag{2}$$

where $\Phi_\rho(z,w)$ is the standard bivariate normal distribution with correlation coefficient ρ, and define

$$H(x,y;\rho) := \Phi_\rho(\Phi^{-1}(F_X(x)), \Phi^{-1}(F_Y(y))), \tag{3}$$

then we have

$$F(x,y) = H(x,y;\rho). \tag{4}$$

Equation (3) is termed as meta-Gaussian distribution of X and Y [1], which is an exact model for $F(x,y)$ under the normality assumption of $P(Z \le z, W \le w)$.

This normality assumption, however, may not hold in certain real-world applications. If that is the case, it seems that what one can hope is that $H(x,y;\rho)$ is still close to $F(x,y)$. This leads to the question of whether we can tune the model to make it closer to $F(x,y)$. In particular, we ask whether an optimal value can be found to replace the ρ-value obtained by transforming X and Y, such that the model is optimal under a given performance criterion. Furthermore, we ask whether $H(x,y;\rho) := \Phi_\rho(\Phi^{-1}(F_X(x)), \Phi^{-1}(F_Y(y)))$ is still a CDF with ρ replaced by an arbitrary value in the unit interval. While the first question may be tackled by conducting numerical experiments, the second question can be answered using the copula theory. Indeed, using the Gaussian copula, we can redefine the BMGD with ρ decoupled from X and Y.

The bivariate Gaussian copula has the following form [19]:

$$C_\gamma(u,v) = \Phi_\gamma(\Phi^{-1}(u), \Phi^{-1}(v)), \tag{5}$$

where Φ_γ is the bivariate standard normal distribution with correlation parameter γ in $(-1,1)$, u and v take values in the unit interval. Writing $u = F_X(x)$ and $v = F_Y(y)$, we recognize that the right-hand side of (3) becomes a Gaussian copula. Conversely, given a bivariate Gaussian copula and two CDFs $F_X(x)$ and $F_Y(y)$, Sklar's theorem [20], entails that

$$H(x,y;\gamma) = \Phi_\gamma(\Phi^{-1}(F_X(x)), \Phi^{-1}(F_Y(y))) \tag{6}$$

is a joint CDF with marginal distributions $F_X(x)$ and $F_Y(y)$. Here, we use γ in the equation to emphasize that its value is not necessarily the same as that of ρ, which is obtained from X and Y. Generally, $H(x,y;\gamma)$ is the CDF of a random vector with X and Y as its marginal variables and a dependence structure that is different than that of $F(x,y)$. Note that $H(x,y;\gamma)$ is not equal to $F(x,y)$

even when $\gamma = \rho$ unless the normality condition given by (2) holds. Here, we denote this random vector by $(X, Y)_\gamma$ and refer to (6) as the meta-Gaussian distribution induced from X and Y. Clearly, the BMGD is simply the bivariate Gaussian copula expressed in the original variables. While $(X, Y)_\gamma$ is an BMGD with an arbitrarily specified γ value in $(-1, 1)$, to obtain an approximate model to (X, Y), the dependence parameter needs to be estimated using some estimator.

Equation (7) below is used in the simulation study of this work. For $(X, Y)_\gamma$, let $H_{Y|X}(y_{p|x}|x)$ denote the conditional distribution of Y given $X = x$. An expression for $H_{Y|X}(y_{p|x}|x)$ is given in Appendix A. For a value of p in $(0, 1)$, the p-probability conditional quantile of Y given $X = x$ is a value $y_{p|x}$ of Y such that $p = H_{Y|X}(y_{p|x}|x)$. From Equation (A2) in Appendix A, we have

$$y_{p|x} = F_Y^{-1}\left(\Phi\left(\gamma\Phi^{-1}(F_X(x)) + (1 - \gamma^2)^{1/2}\Phi^{-1}(p)\right)\right). \tag{7}$$

This equation is also given in [21], obtained by taking partial derivative of the copula with respect to u, wherein the quantity $y_{p|x}$ is referred to as the p-th copula quantile curve of y conditional on x.

2.2. Estimation of the Dependence Parameter

The dependence parameter γ can be estimated once the marginals are specified. This estimation approach exploits the fundamental idea of separation between the univariate marginals and the dependence structure of the copula theory. Three estimation methods are described here.

2.2.1. Pearson's Correlation

Apparently, we can let $\gamma = \rho$, the Pearson's correlation coefficient between the two standard normal variables Z and W, defined in the previous section. In theory, if the normality condition holds, this value renders the BMGD modeling exact.

2.2.2. Maximum Likelihood

The method of maximum-likelihood is commonly used in parameter estimation for a statistical model. The maximum likelihood estimator (MLE) is statistically consistent and asymptotically normally distributed [22]. In this work, the method is used to estimate only the dependence parameter (separated from estimating parameters of the marginals), which can be thought as the second stage of the two-stage procedure for estimating copula parameters [23]. An application of this method in the context of multivariate meta-Gaussian distribution models is given by [24].

The MLE for γ in $H(x, y; \gamma)$ can be obtained by following the standard procedure. Assume we have a sample of n points: $(x_1, y_1), (x_2, y_2), \ldots (x_n, y_n)$. The likelihood function is

$$L(\gamma) = \prod_{i=1}^{n} h(x_i, y_i; \gamma).$$

The corresponding log-likelihood function is

$$l(\gamma) = ln(L(\gamma)) = \sum_{i=1}^{n} ln(h(x_i, y_i; \gamma)).$$

Setting the derivative of the log-likelihood function to zero yields the following cubic equation:

$$\gamma^3 - \left(\frac{1}{n}\sum_{i=1}^{n} z_i w_i\right)\gamma^2 + \left(\frac{1}{n}\sum_{i=1}^{n}(z_i^2 + w_i^2) - 1\right)\gamma - \frac{1}{n}\sum_{i=1}^{n} z_i w_i = 0, \tag{8}$$

where z_i and w_i are related to x_i and y_i, respectively, through NQT described earlier in this section. The MLE is given by the real root of the above equation. Let $b = -\frac{1}{n}\sum_{i=1}^{n} z_i w_i$ and $c = \frac{1}{n}\sum_{i=1}^{n}(z_i^2 +$

$w_i^2) - 1$. Let $B = 2b^3 - 9bc + 27b$, $C = \left(B^2 - 4(b^2 - 3c)^3\right)^{\frac{1}{2}}$, and $D = \left(\frac{1}{2}(C + B)\right)^{\frac{1}{3}}$. The real root of (8) is the following:

$$\gamma = -\frac{1}{3}\left(b + D + \frac{b^2 - 3c}{D}\right). \tag{9}$$

2.2.3. Minimization of the Mallows Distance

The Mallows distance is a similarity measure between two probability distributions. It can be defined in a number of ways at various levels of abstraction. Here, the one in [25] is given. Let P and Q be probability distributions on R^n. Consider joint distributions of the form (X, Y) with X following P and Y following Q. Let F be a joint distribution of X and Y. The Mallows distance between P and Q can be defined by the minimum of the expected difference between X and Y, taken over all joint probability distributions F:

$$M_p(P, Q) = \min_F \left\{ (E\|X - Y\|^p)^{1/p} : (X, Y) \sim F, X \sim P, Y \sim Q \right\}, \tag{10}$$

where $\| \cdot \|$ is usually taken to be the Euclidean or L^1 norm and $p \geq 1$. In this study, we set $p = 1$, in light of the fact that if P and Q are one-dimensional, the Mallows distance is simply the area between the two CDF curves [25]. This particular case is recognized as the area validation metric [26] in the discussion of proper divergence functions for evaluating differences between model output probability distributions and the corresponding empirical distributions of the data. Some important properties of the Mallows distance are given in [25,27].

The Mallows distance is related to another similarity measure known as earth mover's distance (EMD), introduced by [28] in the context of image retrieval. The Mallows distance is equivalent to the EMD when the EMD is applied to probability distributions. The equivalence, on the one hand, provides theoretical justification for the EMD by the theory of the Mallows distance, as shown by [25]; and, on the other hand, offers efficient algorithms available to the EMD for computing the Mallows distance numerically. Thus, optimal dependence parameter values are obtained by minimizing the EMD.

The EMD is a general and flexible metric for evaluating similarity or distance between two multi-dimensional distributions. The EMD is a cross-bin comparison measure. As such, it is more robust than those bin-by-bin comparison measures (histogram matching techniques such as Minkowski-form distance and histogram intersection). Bin-by-bin comparison measures have two major drawbacks: (1) they account only for similarity between corresponding bins; and (2) they are sensitive to bin size and therefore the selection of a proper bin size is important. By contrast, the EMD also uses information across bins and always yields better results with finer bins. For a detailed discussion of the EMD and its properties, see [28].

3. A Bivariate Distribution Model for Precipitation Amounts

Because of the intermittent nature of precipitation [29], modeling statistical relationships among variables of precipitation amounts can be demanding. For a single variable of precipitation amounts, one can employ a mixed-type distribution model that has a point mass at zero and a continuous probability density function over the positive domain. This modeling approach was adapted in [10] to introduce a mixed-type model for the bivariate case. For this mixed-type bivariate distribution, Herr and Krzysztofowicz [8] formulated equations for conditional distributions given a value of a marginal distribution. A full derivation of the equations using Dirac delta functions were given in [30,31]. In this section, an alternative derivation is given using calculus and elementary probability theory. The derivation also provides a proof for Equation (D1) in [5].

Let X and Y denote accumulated precipitation amounts for a given temporal scale. We note here that $X \geq 0$ and $Y \geq 0$. Denote the joint CDF of X and Y by $F(x, y)$. Then, $F(x, y)$ can be expressed as:

$$F(x, y) = p_{00} + p_{10}G_X(x) + p_{01}G_Y(y) + p_{11}D(x, y), \tag{11}$$

where

$$
\begin{aligned}
p_{00} &= P(X = 0, Y = 0), \\
p_{10} &= P(X > 0, Y = 0), \\
p_{01} &= P(X = 0, Y > 0), \\
p_{11} &= P(X > 0, Y > 0), \\
G_X(x) &= P(X \le x | X > 0, Y = 0), \\
G_Y(y) &= P(Y \le y | X = 0, Y > 0), \\
D(x,y) &= P(X \le x, Y \le y | X > 0, Y > 0).
\end{aligned}
$$

Here, $G_X(x)$, $G_Y(y)$, and $D(x,y)$ are assumed to be continuous. In this work, $D(x,y)$ is modeled by the BMGD. It is easy to see that the point probability masses $p_{00}, p_{10}, p_{01}, p_{11}$ sums to 1.

To obtain expressions for the following conditional distribution

$$
F_{Y|X}(y|x) := P(Y \le y | X = x), \tag{12}
$$

we further define:

$$
\begin{aligned}
D_X(x) &:= P(X \le x | X > 0, Y > 0), \\
D_{Y|X}(y|x) &:= P(Y \le y | X = x, X > 0, Y > 0), \\
F_{X|X>0}(x) &:= P(X \le x | X > 0).
\end{aligned}
$$

Then, for $x = 0$, we have

$$
F_{Y|X}(y|x) = a + (1 - a)G_Y(y), \tag{13}
$$

where $a = p_{00}/(p_{00} + p_{01})$. If $p_{00} + p_{01} = 0$, then $P(X = 0) = 0$, indicating that (13) can not be defined for this case. For $x > 0$, we have

$$
F_{Y|X}(y|x) = c(x) + (1 - c(x))D_{Y|X}(y|x), \tag{14}
$$

where

$$
c(x) = \frac{p_{10}g_X(x)}{p_{10}g_X(x) + p_{11}d_X(x)}, \tag{15}
$$

with g_X being the density function corresponding to G_X and d_X corresponding to D_X. If $p_{10} = 0$, Equation (14) reduces to $F_{Y|X}(y|x) = D_{Y|X}(y|x)$. If $p_{10} = 0$ and $p_{11} = 0$, then $P(X > 0) = 0$, implying that the case of $x > 0$ is void.

A proof for (15) using calculus and elementary probability theory is given in Appendix B. A similar form of this equation can be found in [8], which was derived with the use of Dirac delta functions [31].

4. Numerical Experiments

4.1. Data

The idea of tuning the BMGD model is tested using real-world data. The experimental data sets are collected for drainage basins or sub-areas of drainage basins. For the observed variable, historical values of 6-hly mean areal precipitation (MAP) for the drainage areas are used. For the forecast variable, 6-hly precipitation reforecasts generated from the Global Ensemble Forecast System (GEFS) of the U.S. National Centers for Environmental Prediction (NCEP) are used. A more detailed description of the data is given as follows.

The drainage areas considered in this study are selected from the service areas of the Arkansas–Red Basin (AB-), Colorado Basin (CB-), California–Nevada (CN-), and Middle Atlantic (MA-) River Forecast

Centers (RFC) of the US National Weather Service (NWS), which are of different geophysical and climate characteristics [14]. Specifically, four drainage areas are selected as follows: the Chikaskia River at Blackwell in Oklahoma at ABRFC (ID: BLKO2); the middle zone of the Dolores River at Dolores in Colorado at CBRFC (ID: DOLC2LMF); the lower zone of the Eel River at Fort Seward in California at CNRFC (ID: FTSC1LLF); and the West Branch Delaware River at Walton in New York at MARFC (ID: WALN6).

The RFC-provided historical precipitation observations are MAP values accumulated over 6-h time periods. These 6-h MAP values came from rain-gauge only analysis or multisensor analysis [32], recorded in local time. Archives of these historical MAP values are available typically for several decades.

The GEFS version used is based on version 9.0.1 of the NCEP Global Forecast System (GFS), which is a deterministic forecast model. The GEFS uses the GFS model to produce a control member and then perturbs it to generate other ensemble members. This version has a horizontal resolution of T254 ($\sim$55 km) for forecast period of days 1–8 and T190 ($\sim$70 km) for forecast period of days 7.5–16. This GEFS version was used by the Earth System Research Laboratory of the US National Oceanic and Atmospheric Administration to produce GEFS reforecast datasets. The datasets are generated for the period between 1985 and 2010 [33]. Of the various configurations available for the GEFS reforecasts, the one used here has 11 ensemble members, produced daily with a forecast start time at 00 UTC.

Corresponding to a given MAP value, the mean value of the ensemble members (the GEFS reforecast ensemble has 11 members) is used for the forecast variable in the construction of the BMGD model. As a side remark, the practice of employing the ensemble mean in ensemble post-processing has been evaluated in several recent studies, conducted using long reforecast records (see, for example, [14]). The results in [14] indicate that the forecast ensembles generated by the BMGD model typically underestimate the largest observed precipitation amounts, but otherwise are reliable.

To create forecast-observation pairs, the 6-hly MAP values (recorded at hours 0, 6, 12, and 18 in local time) are matched with the GEFS ensemble mean values (issued for hours 0, 6, 12, and 18 in UTC) in such a way that their time differences in UTC are minimum. The GEFS reforecasts are gridded datasets. The reforecasts produced at the grid point that is closest to the centroid of the drainage area is selected. This is a regridding method referred to as the nearest neighbor (NN) method in the literature. More sophisticated regridding methods (e.g., bilinear interpolation) can be found in [34]. Whether another method would outperform the NN method depends on factors including the distance between the centroid and the nearest grid point as well as the geophysical and climate characteristics of the drainage area. Table 1 shows the coordinates of the centroid of the drainage areas and those of the corresponding GEFS grid points. We note that the differences between the centroid coordinates and the corresponding grid point coordinates are in the range of 0.06–0.41 degree for the latitude and 0.06–0.22 degree for the longitude, which are deemed small. Thus, the simple NN method is used instead of a more sophisticated method.

Table 1. Drainage area centroids and corresponding GEFS grid points.

	Centroid		Grid Point	
	Lon ($^\circ$W)	Lat ($^\circ$N)	Lon ($^\circ$W)	Lat ($^\circ$N)
BLKO2	97.28	36.81	97.50	37.22
DOLC2LMF	108.22	37.63	108.28	37.69
FTSC1LLF	123.15	39.60	123.28	39.56
WALN6	75.14	42.17	75.00	42.37

4.2. Simulation

The statistical models and procedures described in the previous sections are coded using the R language to perform numerical simulations. Some R routines used in the simulations are obtained from the following R packages of the Comprehensive R Archive Network:

- Package 'lmomco': Providing extensive functions for computation of L-moments in addition to probability weighted moments, and parameter estimation for numerous distributions.
- Package 'emdist': Providing tools for computing the Earth Mover's Distance.

Generally, precipitation amounts exhibit seasonal variations. Additionally, the forecast skill of the GEFS varies during the year. To account for these seasonal variations, seasonal windows of about 90 days are used to subset the historical MAP data and GEFS reforecast data so that simulations are performed for each of the four seasons: spring (Mar., Apr., May), summer (Jun., Jul., Aug.), fall (Sep., Oct., Nov.), and winter (Dec., Jan, Feb). For a given season, this study is focused on the forecast period of the first 24 h with precipitation amounts aggregated over the four 6-h sub-periods.

The three parameter Weibull distribution is used for modeling the two marginal distributions of the BMGD and the density functions involved in computing the $c(x)$ in Equation (14). Simulated forecasts are generated from the fitted marginal distribution for the forecast. A common rain gauge detection limit is 0.01 inches. This value is used here as the threshold to distinguish between wet and dry conditions. Any MAP values and GEFS reforecast values less than this threshold are set to zero. The choice of the three parameter Weibull in this study is based on goodness-of-fit (GOF) by visual inspection of the CDF and quantile-quantile plots. It is found that the three parameter gamma distribution yields similar results.

The distance between the modeled and the empirical distributions can also be evaluated by comparing their conditional distributions. This way, a particular portion of the BMGD can be assessed for GOF. For example, assessing GOF for conditional distributions given large forecast amounts is important in forecasting heavy precipitation events for flood forecasting. In this experiment, GOF of the conditional distribution of the observed given the forecast greater than the 75th percentile of the forecasts obtained from ensemble mean values of the retrospective GEFS ensemble forecasts is also studied.

The BMGD is dependent on the value of γ. This simulation study is aimed to evaluate the GOF, unconditionally and conditionally, between the empirical joint distribution and the modeled BMGD as γ varies. Optimal γ values may be found when a certain distance measure between the two distributions is minimum. In this study, two distance measures are used: the EMD and the well-known Kolmogorov–Smirnov statistic (KSS). We note here that the KSS is applied only for conditional distributions treated as one-dimensional. For each γ value, the number of simulated forecast-observation pairs is set to 50,000. For a given drainage area and season, a simulation run is executed as follows:

- Loop through a sequence of γ values. These values are created in increments of a certain step size.
- For a given γ value, loop to create simulated forecast-observation pairs.
- Draw a sample point from the forecast distribution.
- If this value is zero, generate an observation sample point from the distribution given in Equation (13). In this equation, the constant a dictates the probabilities of drawing a zero value or a non-zero value from $G_Y(y)$.
- If the simulated forecast value is positive, generate an observation sample point from the distribution given in Equation (14). In this equation, the function $c(x)$ dictates the probabilities of drawing a zero value or a non-zero value from $D_{Y|X}(y|x)$.

For each γ value, a value of a distance measure is computed. This gives us a sequence of distance values. These values fluctuate as a function of γ. To smooth out the fluctuations, a moving average with five steps is applied. The average is taken over the central value and the four values surrounding it.

4.3. Results

We now illustrate that the normality assumption discussed in Section 2 may fail to hold when dealing with real-world problems. Figure 1 shows the scatter plots of the transformed forecast and observed for drainage areas BLKO2 (a) and FTSC1LLF (b) for summer and winter, respectively. In the plots, the Z-axis represents the forecast and the W-axis the observed. We can see that neither scatter plot has an elliptical shape, a characteristic of the bivariate normal distribution. The straight edges seen in the plots can be attributed to the following: (1) the observed and forecast variables are right-skewed with a relatively large probability mass at the threshold value, which is 0.01 inches (a common rain gauge detection limit); and (2) when the observations are just 0.01 inches, the corresponding forecasts are spread out considerably, and vice versa. The lack of normality may also be caused by the rain gauges' resolution: for small rainfall values, a high percentage of identical values are yielded. This may affect the transformed variables through NQT.

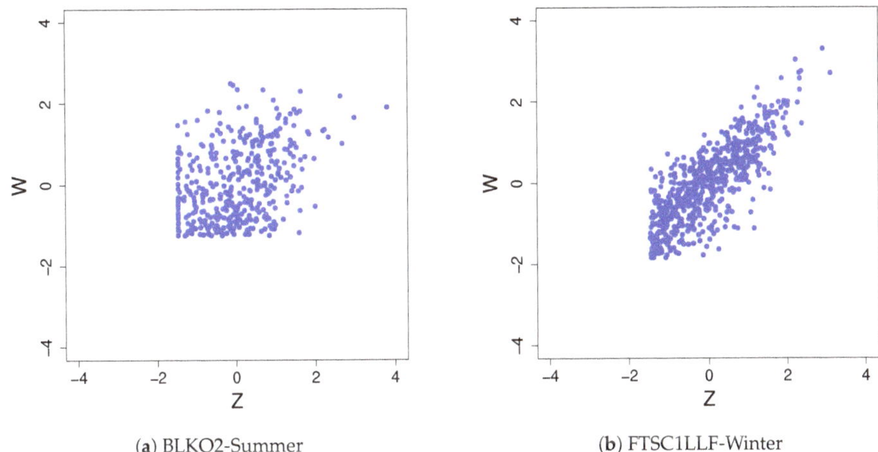

(**a**) BLKO2-Summer   (**b**) FTSC1LLF-Winter

Figure 1. Scatter plots of the transformed forecast and observed.

Tuning the dependence parameter unconditionally is examined next. Figure 2 shows EMD values for a sequence of γ values and contrast them with the EMD values obtained from applying the sample Pearson's correlation coefficient (SPCC), the maximum likelihood estimate (MLE), and the minimum EMD (MEMD), for the following combinations of drainage areas and seasons: DOLC2LMF-summer (a), BLKO2-Fall (b), FTSC1LLF-Winter (c), and WALN6-Summer (d). In the plots, the circles are EMD values at the corresponding γ values. The dots give smoothed EMD values. The solid triangle, diamond, and square marks the smoothed EMD values obtained from applying the SPCC, the MLE, and the MEMD, respectively. We can make a few observations from the plots. Firstly, the solid square (MEMD), as the minimum value of the smoothed EMD, is necessarily located below other EMD values in the plots. Secondly, in panel (a), all three methods yield EMD values that are computationally and statistically indistinguishable. Thirdly, the MEMD γ value is located to the right of the SPCC and MLE values for all four cases. For cases (b), (c), and (d), large percentage reductions in EMD are achieved by the MEMD method in comparison with the other two methods, with the MEMD γ value much greater than that of the other two methods.

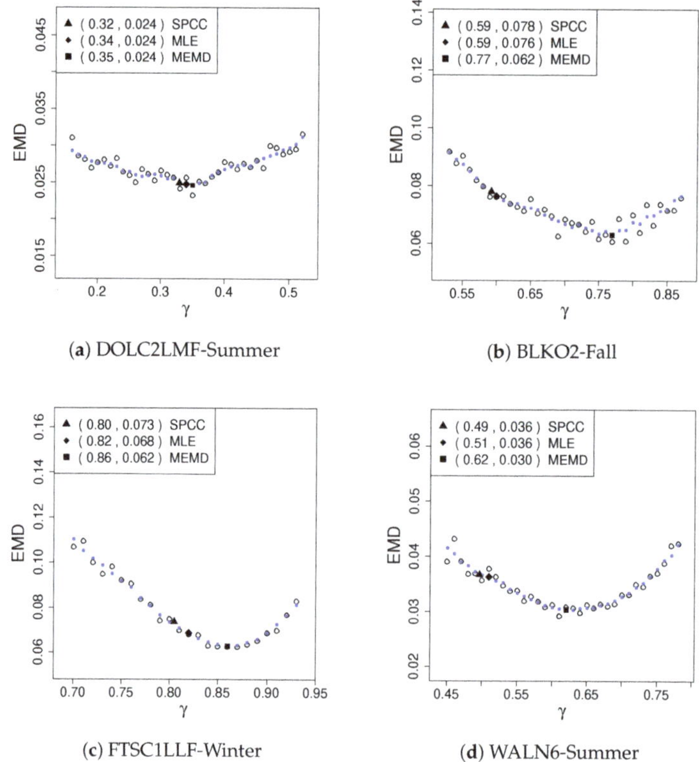

(**a**) DOLC2LMF-Summer

(**b**) BLKO2-Fall

(**c**) FTSC1LLF-Winter

(**d**) WALN6-Summer

Figure 2. EMD values for the methods of SPCC, MLE, and MEMD.

Table 2 presents smoothed EMD values obtained from applying the SPCC, the MLE, and the MEMD for all the drainage areas and seasons considered in this work. We can see that, in the majority of cases, the SPCC and MLE give similar results. For BLKO2, all three methods give close EMD values except for Fall. For DOLC2LMF, the largest difference occurs for Spring where the EMD given by the SPCC is about 17% greater than that given by the MEMD. For FTSC1LLF, relative reduction in EMD ranging from 8% for Fall to about 18% for Winter is achieved by the MEMD. For WALN6, a 20% reduction in EMD occurs for Summer and a 30% reduction for Fall, respectively. For some cases, the results from the three estimation methods are close, statistically indistinguishable. For other cases, significant reductions of 10–20% in EMD are seen by MEMD relative to SPCC or MLE. It is worth noting that, for the case of BLKO2-Summer, despite the obvious lack of normality as indicated by Figure 1a, the MEMD is unable to provide any improvement. The results show that the MEMD performs better only in one-fourth of the cases. Overall, the results indicate that tuning the BMGD unconditionally under the MEMD does not bring much performance improvement over the SPCC and MLE, even though the performance criterion is MEMD itself, which is favorable. This lack of improvement may be explained in the context of conditional tuning. While conditional tuning for the largest quartile (forecast > 75th percentile) is investigated (results will be reported next), conditional tuning for the three smaller quartiles are also examined (results are not presented as the focus of this work is on the largest quartile). There seems to be a tendency that smaller quartiles correspond to smaller conditional tuning γ values. It stands to reason that a γ value yielded by the unconditional tuning represents the middle ground of the conditional tuning γ values. This middle ground value lies close to the SPCC and MLE γ values in the majority of cases.

Table 2. EMD between the empirical bivariate CDF and its simulated CDFs.

		Spring		Summer		Fall		Winter	
		γ	EMD	γ	EMD	γ	EMD	γ	EMD
BLKO2	SPCC	0.62	0.057	0.36	0.049	0.59	0.078	0.77	0.041
	MLE	0.66	0.059	0.41	0.049	0.59	0.076	0.81	0.041
	MEMD	0.62	0.057	0.40	0.049	0.77	0.062	0.78	0.041
DOLC2LMF	SPCC	0.38	0.028	0.32	0.024	0.31	0.030	0.40	0.031
	MLE	0.39	0.026	0.34	0.024	0.35	0.029	0.43	0.029
	MEMD	0.46	0.024	0.35	0.024	0.36	0.028	0.51	0.028
FTSC1LLF	SPCC	0.77	0.049	0.58	0.083	0.77	0.066	0.80	0.073
	MLE	0.79	0.046	0.65	0.076	0.79	0.063	0.82	0.068
	MEMD	0.85	0.042	0.62	0.074	0.87	0.061	0.86	0.062
WALN6	SPCC	0.70	0.032	0.49	0.036	0.72	0.035	0.82	0.035
	MLE	0.72	0.032	0.51	0.036	0.74	0.029	0.84	0.035
	MEMD	0.74	0.031	0.62	0.030	0.77	0.027	0.78	0.033

As discussed previously, there are situations where the conditional distribution of the observed variable given the forecast in a certain range of its domain is of particular interest. In these situations, one may target this specific conditional distribution to achieve a better fit through model tuning, while recognizing that the dependence parameter tuned this way generally does not provide optimal values for the entire joint distribution. Here, the targeted conditional CDFs are the ones obtained with the observed variable conditioned on the corresponding forecast variable being greater than the 75th percentile. This conditional tuning is illustrated by Figure 3, which shows one-dimensional EMD (EMD1d) values for a sequence of γ values for the same combinations of drainage areas and seasons used in Figure 2. In the plots, the circles represent the EMD1d values at the corresponding γ values. The dots represent the smoothed EMD1d values. For all of the cases considered in this work, the optimal γ values for the minimized smoothed EMD1d values are given in Table 3 below. The effect of this localized optimization can be demonstrated by Figure 4. This figure shows one-dimensional conditional CDFs for the observed variable for DOLC2LMF-Spring (a) and FTSC1LLF-Fall (b). The conditional CDFs are obtained with the observed variable conditioned on the corresponding forecast variable being greater than the 75th percentile. In each plot, along with an empirical CDF, CDFs obtained from simulations using the methods of SPCC (sim3-gray), MEMD applied to the bivariate distributions as described above (sim2-red), and MEMD applied to only the conditional distributions (sim1-blue) are given. We can see that sim1 yields the best fit for both cases. Since the EMD measures the area between two one-dimensional CDFs (Section 2), this outcome demonstrates well the fact that minimizing the EMD amounts to minimizing this area.

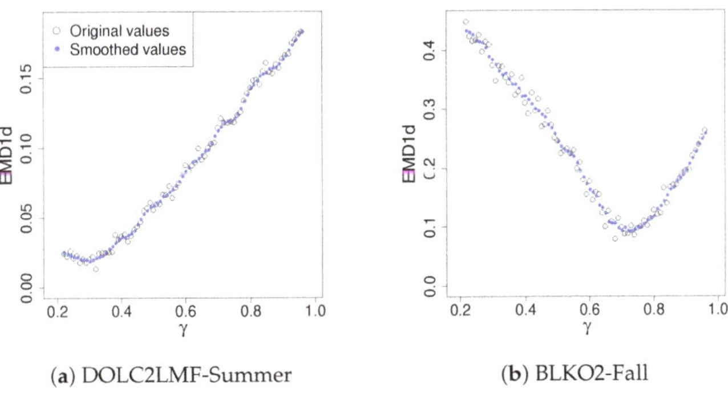

(**a**) DOLC2LMF-Summer   (**b**) BLKO2-Fall

Figure 3. *Cont.*

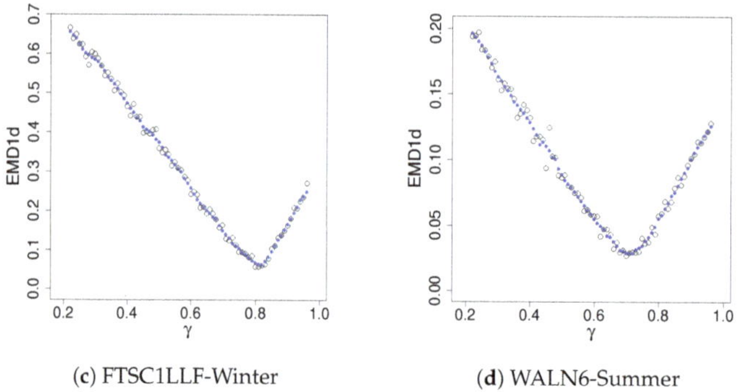

(**c**) FTSC1LLF-Winter

(**d**) WALN6-Summer

Figure 3. EMD values as a function of γ for.

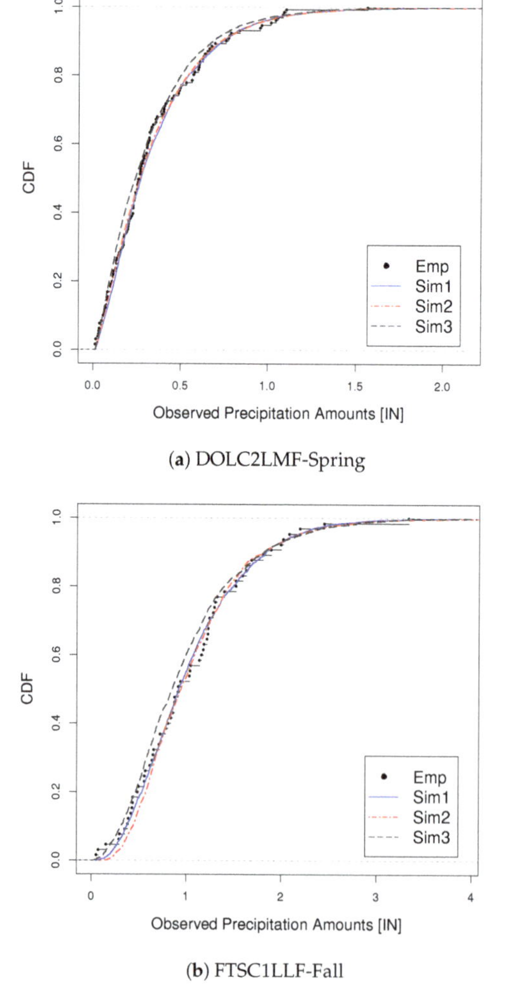

(**a**) DOLC2LMF-Spring

(**b**) FTSC1LLF-Fall

Figure 4. The conditional empirical CDF and its simulated CDFs.

Table 3. Results for conditional tuning.

		Spring			Summer			Fall			Winter		
		γ	EMD1d	KSS	γ	EMD1d	KSS	γ	EMD1d	KSS	γ	EMD1d	KSS
BLKO2	SPCC	0.62	0.069	0.084	0.36	0.072	0.077	0.59	0.169	0.107	0.77	0.102	0.117
	MLE	0.66	0.053	0.067	0.41	0.072	0.071	0.59	0.169	0.107	0.81	0.079	0.094
	MEMD2d	0.62	0.069	0.084	0.40	0.070	0.070	0.77	0.108	0.126	0.78	0.095	0.112
	MEMD1d	0.68	0.049	0.068	0.39	0.069	0.070	0.72	0.090	0.113	0.89	0.072	0.087
	MKSS	0.66	0.053	0.067	0.38	0.070	0.068	0.67	0.119	0.075	0.87	0.076	0.083
DOLC2LMF	SPCC	0.38	0.031	0.096	0.32	0.018	0.066	0.31	0.037	0.089	0.40	0.055	0.087
	MLE	0.39	0.029	0.094	0.34	0.021	0.074	0.35	0.033	0.072	0.43	0.048	0.094
	MEMD2d	0.46	0.018	0.061	0.35	0.023	0.076	0.36	0.033	0.067	0.51	0.042	0.141
	MEMD1d	0.46	0.018	0.059	0.28	0.018	0.053	0.40	0.029	0.066	0.51	0.042	0.120
	MKSS	0.49	0.019	0.058	0.29	0.018	0.053	0.37	0.032	0.063	0.39	0.055	0.085
FTSC1LLF	SPCC	0.77	0.108	0.097	0.58	0.110	0.177	0.77	0.083	0.109	0.80	0.067	0.042
	MLE	0.79	0.096	0.094	0.65	0.096	0.150	0.79	0.069	0.104	0.82	0.066	0.047
	MEMD2d	0.85	0.077	0.081	0.62	0.104	0.149	0.87	0.058	0.090	0.86	0.102	0.071
	MEMD1d	0.84	0.074	0.085	0.70	0.087	0.159	0.84	0.052	0.092	0.82	0.066	0.044
	MKSS	0.84	0.074	0.079	0.63	0.102	0.147	0.85	0.054	0.088	0.80	0.067	0.042
WALN6	SPCC	0.70	0.067	0.099	0.49	0.094	0.164	0.72	0.076	0.127	0.82	0.054	0.161
	MLE	0.72	0.060	0.094	0.51	0.088	0.154	0.74	0.073	0.113	0.84	0.049	0.148
	MEMD2d	0.74	0.053	0.094	0.62	0.044	0.088	0.77	0.062	0.093	0.78	0.067	0.197
	MEMD1d	0.84	0.035	0.077	0.70	0.026	0.066	0.80	0.059	0.082	0.94	0.023	0.067
	MKSS	0.85	0.036	0.073	0.68	0.028	0.064	0.81	0.062	0.075	0.94	0.023	0.063

Table 3 presents more results for tuning the BMGD for the one-dimensional conditional distributions given forecast values being greater than the 75th percentile. In the table, the γ values are obtained from the SPCC, MLE, MEMD2d (MEMD for the BMGD unconditionally), MEMD1d (MEMD for the conditional distribution), and MKSS (the minimum KSS for the conditional distribution). The associated EMD1d and KSS values are smoothed values. While the results are mixed and may seem perplexing, we highlight here the performance of the conditional tuning methods. For WALN6, comparing MEMD1d and MKSS with SPCC, MLE, and MEMD2d, we can see large percentage reductions in EMD1d and KSS for the Spring, Summer, and Winter. For other locations, the results are mixed in a complicated way. To have an objective assessment of the results, we count the number of cases that one method outperforms another. Specifically, to compare a given pair of methods, e.g., MKSS and SPCC, we form a ratio of the number of cases that MKSS outperforms SPCC to the number of cases the MKSS underperforms the SPCC. The performance ratios are tabulated in Table 4. For MKSS vs. SPCC, the ratio in EMD1d is found to be 13:0 in the row with MKSS and SPCC appearing under EMD1d. We can see that the unconditional tuning method MEMD2d outperforms the SPCC considerably as measured by the ratios, but not so much for MEMD2d vs. MLE. The effect of the conditional tuning is much greater than that of the unconditional tuning as the ratios for the MEMD1d and MKSS indicate. Of course, when the choice of tuning (optimization) criterion is EMD1d, it is necessary that the MEMD1d outperforms the other methods. This is true also for MKSS. What is interesting here is that the ratios under KSS compellingly favor the MEMD1d and those under EMD1d favor the MKSS. The results suggest that, when the dependence parameter is optimized for a targeted conditional distribution, the GOF of the conditional distribution can be improved.

Table 4. Performance ratios.

		EMD1d	KSS
MEMD2d	SPCC	12:3	11:4
	MLE	9:6	8:7
MEMD1d	SPCC	15:0	13:3
	MLE	15:0	12:4
	MEMD2d	14:0	12:3
MKSS	SPCC	13:0	15:0
	MLE	12:3	15:0
	MEMD2d	11:3	16:0

5. Discussion and Concluding Remarks

Precipitation is intermittent. In order to model the joint distribution between two variables of precipitation amounts, a mixed-type bivariate distribution is considered in this paper. A new proof for an intermittency equation of the model is given. With the use of an embedded bivariate meta-Gaussian distribution, this modeling approach allows for an explicit treatment of precipitation intermittency as well as a wide choice of parametric and non-parametric models for the marginal distributions. The focus of this work is on the question of whether and how the bivariate meta-Gaussian distribution can be tuned to yield an optimal fit. The meta-Gaussian distribution can be formulated in two different ways. In the conventional formulation, a normality assumption is needed. In certain real-world applications, however, this assumption may not hold. The second formulation, starting from a copula setup, forgoes this assumption and thus allows for model tuning. In this study, predicted single-valued precipitation amounts for a river basin area constitute one variable, and the corresponding observed precipitation amounts the other. The study of this modeling approach is motivated by a need to quantify the uncertainty in the single-valued precipitation forecasts in certain hydrometeorologic applications. In this setting, the uncertainty can be quantified by the conditional distributions given the forecasts.

Numerical simulations are carried out using real-world data from the Global Ensemble Forecast System of the U.S. National Centers for Environmental Prediction and four U.S. National Weather Service River Forecast Centers. The parameter in the meta-Gaussian distribution that characterizes the dependence of the two marginals is tuned under the Mallows distance for the entire joint distribution to yield an optimal value. The results obtained from this optimization are compared with the results obtained from using the sample correlation coefficient and maximum likelihood estimate as parameter values. It is found that the sample correlation coefficient and maximum likelihood estimate produce similar results for all the cases studied. Additionally, for three-fourths of the cases, results yielded by the sample correlation coefficient and the minimum Mallows distance are very close. These results suggest that tuning this dependence parameter has limited effects in altering the behaviour of the meta-Gaussian distribution model toward a better overall model fit.

It is known that the statistical relationship between a precipitation forecast variable and its corresponding observed variable can be nonlinear and heteroscedastic. Specifically, heteroscedasticity here refers to the situation in which the variance of the observed precipitation amounts changes over the range of the forecast precipitation amounts. The dependence parameter values determined from the conventional methods can be thought of as a middle-ground for balancing the fit across this range. When the meta-Gaussian distribution model for precipitation amounts is applied in predicting floods, goodness-of-fit (GOF) for the heavy precipitation amounts in the modeling has a larger impact than that for the light precipitation amounts, which entails a dependence parameter value tuned for the heavy precipitation amounts. This important aspect is investigated by applying the Mallows distance and the Kolmogorov–Smirnov statistic to finding an optimal dependence parameter value for the conditional distribution given the forecast greater than the 75th percentile of the forecast values.

The results suggest that, when the dependence parameter is optimized for a targeted conditional distribution, the GOF of the conditional distribution can be improved.

Funding: This work was partially funded by the Advanced Hydrologic Prediction Service (AHPS) program of the National Weather Service (NWS), and by the Climate Predictions Program for the Americas (CPPA) of the Climate Program Office (CPO), both from the National Oceanic and Atmospheric Administration (NOAA). This support is gratefully acknowledged.

Acknowledgments: The author would like to thank the editors and referees for their valuable comments that lead to improved content and clarity of this work. Special thanks go to Yu Zhang of the NWS Office of Prediction for helpful suggestions.

Conflicts of Interest: The author declares no conflict of interest. The funders had no role in the design of the study; in the collection, analyses, or interpretation of data; in the writing of the manuscript, and in the decision to publish the results.

Appendix A

Equation (7) is used in the simulation study of this work. A derivation of this equation is included below for completeness. The derivation is different than the one given in [35]. We start with the density function of (6), denoted here by $h(x, y; \gamma)$:

$$h(x, y; \gamma) = f_X(x) f_Y(y) \frac{\phi_\gamma(\Phi^{-1}(F_X(x)), \Phi^{-1}(F_Y(y)))}{\phi(\Phi^{-1}(F_X(x)))\phi(\Phi^{-1}(F_Y(y)))}, \tag{A1}$$

where f_X, f_Y, ϕ, and ϕ_γ are the density functions corresponding to F_X, F_Y, Φ, and Φ_γ, respectively. From Equation (6), by definition, we have:

$$
\begin{aligned}
& H(x, y; \gamma) \\
={} & \int_{-\infty}^{\Phi^{-1}(F_X(x))} \left(\int_{-\infty}^{\Phi^{-1}(F_Y(y))} \phi_\gamma(s, t)\, dt \right) ds \\
={} & \int_{-\infty}^{\Phi^{-1}(F_X(x))} \left(\int_0^{F_Y(y)} \phi_\gamma(s, \Phi^{-1}(z)) \frac{1}{\phi(\Phi^{-1}(z))}\, dz \right) ds \\
={} & \int_{-\infty}^{\Phi^{-1}(F_X(x))} \left(\int_{-\infty}^{y} \phi_\gamma(s, \Phi^{-1}(F_Y(t))) \frac{f_Y(t)}{\phi(\Phi^{-1}(F_Y(t)))}\, dt \right) ds \\
={} & \int_{-\infty}^{x} \int_{-\infty}^{y} \phi_\gamma(\Phi^{-1}(F_X(s)), \Phi^{-1}(F_Y(t))) \frac{f_X(s) f_Y(t)}{\phi(\Phi^{-1}(F_X(s)))\phi(\Phi^{-1}(F_Y(t)))}\, ds dt.
\end{aligned}
$$

It follows that $H(x, y; \gamma)$ has the density given in (A1).

Using (A1), we can obtain, for $(X, Y)_\gamma$, the conditional distribution of Y given $X = x$:

$$H_{Y|X}(y|x; \gamma) = \Phi \left(\frac{\Phi^{-1}(F_Y(y)) - \gamma \Phi^{-1}(F_X(x))}{(1 - \gamma^2)^{1/2}} \right). \tag{A2}$$

Indeed, from density function (A1), by definition, we have:

$$H_{Y|X}(y|x;\gamma)$$

$$= \int_{-\infty}^{y} \frac{h(x,t;\gamma)}{f_X(x)}\,dt$$

$$= \int_{-\infty}^{y} \phi_{\gamma}(z(x),w(t))\,\frac{f_Y(t)}{\phi(z(x))\phi(w(t))}\,dt$$

$$= \int_{0}^{F_Y(y)} \phi_{\gamma}(z(x),\Phi^{-1}(s))\,\frac{1}{\phi(z(x))\phi(\Phi^{-1}(s))}\,ds$$

$$= \int_{-\infty}^{\Phi^{-1}(F_Y(y))} \phi_{\gamma}(z(x),t)\,\frac{1}{\phi(z(x))}\,dt$$

$$= \int_{-\infty}^{\Phi^{-1}(F_Y(y))} \frac{1}{\sqrt{2\pi}(1-\gamma^2)^{1/2}}\,\exp\left(\frac{-(\gamma z(x)-t)^2}{2(1-\gamma^2)}\right)\,dt$$

$$= \int_{-\infty}^{(\Phi^{-1}(F_Y(y))-\gamma z(x))/\sqrt{(1-\gamma^2)}} \frac{1}{\sqrt{2\pi}}\,\exp\left(-s^2/2\right)\,ds$$

$$= \Phi\left(\frac{\Phi^{-1}(F_Y(y))-\gamma\Phi^{-1}(F_X(x))}{(1-\gamma^2)^{1/2}}\right).$$

Equation (A2) is of the same form as given in [2], although derived using a different approach.

Appendix B

A derivation of Equation (15) using calculus and elementary probability theory is given here. Assume $x > 0$. The conditional distribution $F_{Y|X}(y|x)$ is, by definition, the following limit:

$$F_{Y|X}(y|x) = \lim_{\epsilon \to 0^+} P(Y \leq y \,|\, X \in I_x(\epsilon)), \tag{A3}$$

where $I_x(\epsilon) = (x-\epsilon, x+\epsilon]$, $x - \epsilon > 0$. Now $P(Y \leq y \,|\, X \in I_x(\epsilon))$ can be decomposed as:

$$P(Y \leq y \,|\, X \in I_x(\epsilon)) \tag{A4}$$
$$= P(Y = 0 \,|\, X \in I_x(\epsilon)) + P(0 < Y \leq y \,|\, X \in I_x(\epsilon)).$$

Using Bayes' Theorem, the second term of the right-hand side of this equation can be expressed as

$$P(0 < Y \leq y \,|\, X \in I_x(\epsilon)) \tag{A5}$$
$$= P(Y \leq y \,|\, X \in I_x(\epsilon), Y > 0)P(Y > 0 \,|\, X \in I_x(\epsilon))$$
$$= P(Y \leq y \,|\, X \in I_x(\epsilon), Y > 0)(1 - P(Y = 0 \,|\, X \in I_x(\epsilon))).$$

It follows that (A4) can be written as

$$P(Y \leq y \,|\, X \in I_x(\epsilon)) \tag{A6}$$
$$= P(Y = 0 \,|\, X \in I_x(\epsilon)) +$$
$$(1 - P(Y = 0 \,|\, X \in I_x(\epsilon))P(Y \leq y \,|\, X \in I_x(\epsilon), Y > 0).$$

Next, we show that

$$\lim_{\epsilon \to 0^+} P(Y = 0 \,|\, X \in I_x(\epsilon)) = c(x), \tag{A7}$$

where $c(x)$ is the expression given in (15). By Bayes' Theorem, we have

$$P(Y = 0 \,|\, X \in I_x(\epsilon)) = \frac{P(X \in I_x(\epsilon) \,|\, Y = 0)P(Y = 0)}{P(X \in I_x(\epsilon))}. \tag{A8}$$

The denominator of the right-hand side of (A8) can be written as

$$P(X \in I_x(\epsilon)) = P(X \le x + \epsilon) - P(X \le x - \epsilon).$$

Applying the steps used to obtain (A4)–(A6) to the first term and then the second term of the above equation, we have

$$
\begin{aligned}
& P(X \in I_x(\epsilon)) \\
= & (1 - P(X = 0))(P(X \le x + \epsilon | X > 0) - P(X \le x - \epsilon | X > 0)) \\
= & (1 - P(X = 0)) \int_{x-\epsilon}^{x+\epsilon} f_{X|X>0}(t)\, dt \\
= & 2\epsilon f_{X|X>0}(\hat{x}_\epsilon)(1 - P(X = 0)),
\end{aligned}
$$

where $f_{X|X>0}$ is the density function associated with $F_{X|X>0}$ defined in Section 3, and $\hat{x}_\epsilon$ is a point in $(x - \epsilon, x + \epsilon)$ (First, mean value theorem for integration). Similarly, we can obtain the following equation for $P(X \in I_x(\epsilon) \,|\, Y = 0)$ in (A8):

$$P(X \in I_x(\epsilon) \,|\, Y = 0) = 2\epsilon g_X(\tilde{x}_\epsilon)(1 - P(X = 0 \,|\, Y = 0)),$$

where g_X is the density corresponding to G_X defined in Section 3 and $\tilde{x}_\epsilon$ is in $(x - \epsilon, x + \epsilon)$. It follows that

$$
\begin{aligned}
& \lim_{\epsilon \to 0^+} \left(P(Y = 0 \,|\, X \in I_x(\epsilon)) \right) && \text{(A9)} \\
= & \frac{(1 - P(X = 0 \,|\, Y = 0))g_X(x)}{(1 - P(X = 0))f_{X|X>0}(x)} P(Y = 0) \\
= & \frac{p_{10} g_X(x)}{(p_{10} + p_{11})f_{X|X>0}(x)}.
\end{aligned}
$$

To obtain $c(x)$, it remains to show that

$$(p_{10} + p_{11})f_{X|X>0}(x) = p_{10} g_X(x) + p_{11} d_X(x), \tag{A10}$$

where $d_X(x)$ is the density corresponding to $D_X(x)$ defined in Section 3. Indeed, we have

$$
\begin{aligned}
& (p_{10} + p_{11})F_{X|X>0}(x) \\
= & (p_{10} + p_{11})P(X \le x \,|\, X > 0) \\
= & (p_{10} + p_{11})P(X \le x, X > 0)/P(X > 0) \\
= & P(X \le x, X > 0, Y = 0) + P(X \le x, X > 0, Y > 0) \\
= & p_{10}P(X \le x \,|\, X > 0, Y = 0) + p_{11}P(X \le x \,|\, X > 0, Y > 0) \\
= & p_{10}G_X(x) + p_{11}D_X(x).
\end{aligned}
$$

Now, we assume that $g_X(x)$, $d_X(x)$, and $f_{X|X>0}(x)$ are continuous. Then, the above equation is differentiable on both sides, which implies (A10).

References

1. Kelly, K.S.; Krzysztofowicz, R. A bivariate meta-Gaussian density for use in hydrology. *Stoch. Hydrol. Hydraul.* **1997**, *11*, 17–31. [CrossRef]
2. Krzysztofowicz, R.; Kelly, K.S. *A Meta-Gaussian Distribution with Specified Marginals*; Technical Report; University of Virginia: Charlottesville, VA, USA, 1996.

3.	Li, W.; Duan, Q.; Ye, A.; Miao, C. An improved meta-Gaussian distribution model for post-processing of precipitation forecasts by censored maximum likelihood estimation. *J. Hydrol.* **2019**, *574*, 801–810. [CrossRef]

4.	Schaake, J.C.; Demargne, J.; Hartman, R.; Mullusky, M.; Welles, E.; Wu, L.; Herr, H.; Fan, X.; Seo, D.-J. Precipitation and temperature ensemble forecasts from single-value forecasts. *Hydrol. Earth Syst. Sci. Discuss.* **2007**, *4*, 655–717. [CrossRef]

5.	Wu, L.; Seo, D.-J.; Demargne, J.; Brown, J.D.; Cong, S.; Schaake, J.C. Generation of ensemble precipitation forecast from single-valued quantitative precipitation forecast for hydrologic ensemble prediction. *J. Hydrol.* **2011**, *399*, 281–298. [CrossRef]

6.	Ye, A.; Deng, X.; Ma, F.; Duan, Q.; Zhou, Z.; Du, C. Integrating weather and climate predictions for seamless hydrologic ensemble forecasting: A case study in the Yalong river basin. *J. Hydrol.* **2017**, *547*, 196–207. [CrossRef]

7.	Papalexiou, S.M.; Koutsoyiannis, D. Entropy based derivation of probability distributions: A case study to daily rainfall. *Adv. Water Resour.* **2012**, *45*, 51–57. [CrossRef]

8.	Herr, H.D.; Krzysztofowicz, R. Generic probability distribution of rainfall in space: The bivariate model. *J. Hydrol.* **2005**, *306*, 234–263. [CrossRef]

9.	Serinaldi, F. Copula-based mixed models for bivariate rainfall data: An empirical study in regression perspective. *Stoch Env. Res Risk Assess* **2009**, *23*, 677–693. [CrossRef]

10.	Shimizu, K. A bivariate mixed lognormal distribution with an analysis of rainfall data. *J. Appl. Meteorol.* **1993**, *32*, 161–171. [CrossRef]

11.	Kedem, B.; Chiu, L.S.; North, G.R. Estimation of mean rain rate: Application to satellite observations. *J. Geophys. Res.* **1990**, *95*, 1965–1972. [CrossRef]

12.	Bowers, M.C.; Tung, W.M.; Gao, J.B. On the distributions of seasonal river flows: Lognormal or power law? *Water Resour. Res.* **2012**, *48*, W05536. [CrossRef]

13.	Demargne, J.; Wu, L.; Regonda, S.K.; Brown, J.D.; Haksu, L.; He, M.; Seo, D.-J.; Hartman, R.; Herr, H.D.; Fresch, M.; et al. The Science of NOAA's Operational Hydrologic Ensemble Forecast Service. *Bull. Am. Meteor. Soc.* **2014**, *95*, 79–98. [CrossRef]

14.	Brown, J.D.; Wu, L.; He, M.; Regonda, S.; Haksu, L.; Seo, D.-J. Verification of temperature, precipitation, and streamflow forecasts from the NOAA/NWS Hydrologic Ensemble Forecast Service (HEFS): 1. Experimental design and forcing verification. *J. Hydrol.* **2014**, *519*, 2869–2889. [CrossRef]

15.	Brown, J.D.; He, M.; Regonda, S.; Wu, L.; Haksu, L.; Seo, D.-J. Verification of temperature, precipitation, and streamflow forecasts from the NOAA/NWS Hydrologic Ensemble Forecast Service (HEFS): 2. Streamflow verification. *J. Hydrol.* **2014**, *519*, 2847–2868. [CrossRef]

16.	Fang, H.-B.; Fang, K.-T.; Kotz, S. The meta-elliptical distributions with given marginals. *J. Multivar. Anal.* **2002**, *82*, 1–16. [CrossRef]

17.	Genest, C.; Favre, A.-C.; Beliveau, J.; Jacques, C. Metaelliptical copulas and their use in frequency analysis of multivariate hydrological data. *Water Resour. Res.* **2007**, *43*, W09401. [CrossRef]

18.	Jolliffe, I.T.; Stephenson, D.B. Forecast Verification: A Practitioner's Guide in Atmospheric Science. In *Atmospheric Science*; Jolliffe, I.T., Stephenson, D.B., Eds.; John Wiley and Sons: Chichester, UK, 2003.

19.	Meyer, C. The Bivariate Normal Copula. *Commun. Stat. Theory Methods* **2013**, *42*, 2402–2422. [CrossRef]

20.	Sklar, A. Random Variables, Joint Distribution Functions, and Copulas. *Kybernetika* **1973**, *9*, 449–460.

21.	Bouyé, E.; Salmon, M. Dynamic copula quantile regressions and tail area dynamic dependence in Forex markets. *Eur. J. Financ.* **2009**, *15*, 721–750. [CrossRef]

22.	Rohatgi, V.K.; Ehsanes Saleh, A.K.M. Section 8.7, Maximum likelihood estimators, Theorem 4. In *An Introduction to Probability Theory and Mathematics*, 2nd ed.; John Wiley and Sons: Hoboken, NJ, USA, 1976.

23.	Joe, H. Asymptotic efficiency of the two-stage estimation method for copula-based models. *J. Multivar. Anal.* **2005**, *94*, 401–419. [CrossRef]

24.	Storvik, B.; Storvik, G.; Fjortoft, R. On the Combination of Multisensor Data Using Meta-Gaussian Distributions. *IEEE Trans. Geosci. Remote Sens.* **2009**, *47*, 2372–2379. [CrossRef]

25.	Levina, E.; Bickel, P.J. The Earth Mover's Distance is the Mallows Distance: Some Insights from Statistics. In Proceedings of the Eighth IEEE International Conference on Computer Vision, ICCV 2001, Vancouver, BC, Canada, 7–14 July 2001; pp. 251–256.

26.	Thorarinsdottir, T.L.; Gneiting, T.; Gissibl, N. Using Proper Divergence Functions to Evaluate Climate Models. *SIAM/ASA J. Uncertain. Quantif.* **2013**, *1*, 522–534. [CrossRef]

27. Zhou, D.; Shi, T. Statistical inference based on distances between empirical distributions with applications to airslevel-3 data. In Proceedings of the 2011 NASA Conference on Intelligent Data Understanding, Mountain View, CA, USA, 19–21 October 2011; pp. 129–143.
28. Rubner, Y.; Tomasi, C.; Guibas, L.J. The Earth Mover's Distance as a Metric for Image Retrieval. *Int. J. Comput. Vis.* **2000**, *40*, 99–121. [CrossRef]
29. Trenberth, K.E.; Zhang, Y.; Gehne, M. Intermittency in Precipitation: Duration, Frequency, Intensity, and Amounts Using Hourly Data. *J. Hydrometeorol.* **2017**, *18*, 1393–1412. [CrossRef]
30. Li, C.; Singh, V.P.; Mishra, A.K. A bivariate mixed distribution with a heavy-tailed component and its application to single-site daily rainfall simulation. *Water Resour. Res.* **2013**, *49*, 767–789. [CrossRef]
31. Herr, H.D. A Bivariate Precipitation Uncertainty Processor for Probabilistic River Forecasting. Master's Thesis, University of Virginia, Charlottesville, VA, USA, 1999; pp. 24–31.
32. Seo, D.-J.; Breidenbach, J.P. Real-time correction of spatially nonuniform bias in radar rainfall data using rain gauge measurements. *J. Hydrometeorol.* **2002**, *3*, 93–111. [CrossRef]
33. Hamill, T.M.; Bates, G.T.; Whitaker, J.S.; Murray, D.R.; Fiorino, M.; Galarneau, T.J.; Zhu, Y.; Lapenta, W. NOAA's SecondGeneration Global MediumRange Ensemble Forecast Dataset. *Bull. Am. Meteor. Soc.* **2013**, *94*, 1553–1565. [CrossRef]
34. Jones, P.W. First- and second-order conservative remapping schemes for grids in spherical coordinates. *Mon. Weather Rev.* **1999**, *127*, 2204–2210. [CrossRef]
35. Li, C.; Singh, V.P.; Mishra, A.K. Simulation of the entire range of daily precipitation using a hybrid probability distribution. *Water Resour. Res.* **2012**, *48*, W03521. [CrossRef]

MDPI
St. Alban-Anlage 66
4052 Basel
Switzerland
Tel. +41 61 683 77 34
Fax +41 61 302 89 18
www.mdpi.com

Forecasting Editorial Office
E-mail: forecasting@mdpi.com
www.mdpi.com/journal/forecasting